Adsorption and Transport
at the Nanoscale

Adsorption and Transport at the Nanoscale

Edited by
Nick Quirke

Taylor & Francis
Taylor & Francis Group
Boca Raton   London   New York

A CRC title, part of the Taylor & Francis imprint, a member of the
Taylor & Francis Group, the academic division of T&F Informa plc.

Published in 2006 by
CRC Press
Taylor & Francis Group
6000 Broken Sound Parkway NW, Suite 300
Boca Raton, FL 33487-2742

© 2006 by Taylor & Francis Group, LLC
CRC Press is an imprint of Taylor & Francis Group

No claim to original U.S. Government works
Printed in the United States of America on acid-free paper
10 9 8 7 6 5 4 3 2 1

International Standard Book Number-10: 0-415-32701-6 (Hardcover)
International Standard Book Number-13: 978-0-415-32701-5 (Hardcover)
Library of Congress Card Number 2005051087

Library of Congress Cataloging-in-Publication Data

Quirke, N. (Nick)
    Molecular simulation of adsorption phenomena / Nick Quirke, David Nicholson.
        p. cm.
    Includes bibliographical references and index.
    ISBN 0-415-32701-6 (alk. paper)
    1. Porous materials. 2. Adsorption--Mathematical models. I. Nicholson, D. (David) II. Title.

TA418.9.P^Q85 2005
660'.284235--dc22   2005051087

Visit the Taylor & Francis Web site at
http://www.taylorandfrancis.com

and the CRC Press Web site at
http://www.crcpress.com

Preface

Materials with nanoporous surfaces are used widely in industry as adsorbents, particularly for applications where selective adsorption of one fluid component from a mixture is important. Nanostructured materials are also of current interest for use in nanofluidics devices. In view of their past, current and potential future importance it seems timely to collect together key articles on relevant aspects of computational methods and applications which underpin progress in this field.

Nick Quirke
Imperial College
London
2005

Contributors

A. Boutin
Physical Chemistry Laboratory
University of Paris-Sud
Orsay, France

S. Buttefey
Physical Chemistry Laboratory
University of Paris-Sud
Orsay, France

A.K. Cheetham
International Center for Materials
  Research
University of California
Santa Barbara, California

B. Coasne
Laboratoire de Physicochimie de la
  Matière Condensée
UMR 5617 CNRS & Université de
  Montpellier II
Montpellier, France

A.H. Fuchs
Physical Chemistry Laboratory
University of Paris-Sud
Orsay, France

K.E. Gubbins
North Carolina State University
Department of Chemical
  Engineering
Raleigh, NC

K.-R. Ha
Department of Chemical Engineering
Keimyung University
Taegu, Korea

S.-C. Kim
Department of Physics
Andong National University
Andong, Korea

P. E. Levitz
Laboratoire de Physique de la
  Matière Condensée
CNRS-Ecole Polytechnique
Palaiseau, France

J.M.D. MacElroy
Department of Chemical Engineering
University College Dublin
Dublin, Ireland

D. Nicholson
Department of Chemistry
Imperial College
London, U.K.

G. K. Papadopoulos
Materials Science & Engineering
School of Chemical Engineering
National Technical University
  of Athens
Athens, Greece

J.-W. Park
Department of Chemical
  Engineering
Keimyung University
Taegu, Korea

R. J.-M. Pellenq
Centre de Recherche en Matière
  Condensée et Nanosciences
Marseille, France

N. Quirke
Department of Chemistry
Imperial College
London, U.K.

R. Radhakrishnan
Department of Bioengineering
University of Pennsylvania
Philadelphia, Pennsylvania

S. Samios
Department of Engineering and
  Management of Energy Resources
University of Western Macedonia
Kozani, Greece

F.R. Siperstein
Departament d'Enginyeria
  Quimica, ETSEQ
Universitat Rovira i Virgili
Tarragona, Spain

T. A. Steriotis
Institute of Physical Chemistry
NCSR "Demokritos"
Attikis, Greece

M. Śliwinska-Bartkowiak
Institute of Physics
Adam Mickiewicz University
Poznan, Poland

A. K. Stubos
Institute of Nuclear Technology
  & Radiation Protection
NCSR "Demokritos"
Attikis, Greece

S.-H. Suh
Department of Chemical
  Engineering
Taegu, Korea

M.B. Sweatman
Department of Chemical and
  Process Engineering
University of Strathclyde
Glasgow, Scotland

K.P. Travis
Department of Engineering
  Materials
University of Sheffield
Sheffield, U.K.

Contents

chapter one

Adsorption and transport at the nanoscale

D. Nicholson
N. Quirke
Imperial College

Contents

1.1 Adsorption and characterisation

Materials with amorphous nanoporous surfaces are used widely in industry as adsorbents, particularly for applications where selective adsorption of one fluid component from a mixture is important. Some materials, such as high surface area carbons, are quoted [1] to have a (BET) surface area higher than 3000 m²/ g. Analysis of adsorption isotherms for such materials by molecular based methods reveals that much of the nanoporosity is on the scale of 3 nm or less. Similarly zeolites and clays may have pore sizes typically 1 nm or less, while single walled nanotubes can have sub nanometre inner diameters. At this scale adsorption is dramatically influenced by nanoscale (surface) geometry, by molecular size and the influence of both on cohesive energies. Naturally, to model adsorption in nanoporous materials, a successful approach will require an atomic model of the surface and a molecular theory of adsorption equilibrium.

The characterisation of (nano) porous materials has engaged the attention of a large community of physical scientists over several decades. For example a series of tri-annual meetings* has been devoted solely to this

* http://www.cops7.up.univ-mrs.fr/cops7/

topic since the early 70's, and characterisation generally forms an important section in many other meeting series centred around problems related to adsorption and adsorbents amongst which may be mentioned the Fundamentals of Adsorption series* started in 1983, a series of meetings on the theme of heterogeneity in adsorption,** held in Poland since 1992, and a more recent series on porous materials held in Princeton*** and in the Pacific Basin countries.****

It is reasonable to ask — why is this not a solved problem? Part of the answer lies in the wide and ever increasing variety of porous materials, part in the randomness of many materials and part in the subtlety and complexity of matter at the nanoscale. Moreover, porous materials find many applications in industry, as well as being ubiquitous in the natural world, so there is a strong motivation behind the sustained interest they evoke.

The adsorption of nitrogen at liquid nitrogen temperature was a well established technique by the early years of the last century, and has continued to dominate the scene as a standard method of characterisation, encouraged by the more recent ready availability of fully automated equipment [1]. It forms the basis of many thousands of surface area measurements (via the BET method) and pore size distribution determinations carried out daily in industrial laboratories worldwide.

The interpretation of adsorption data has benefited greatly from simulation studies (methodological details are given in Sweatman and Quirke, this volume), although this has continued to be largely centred on underlying conceptual models of cylindrical pores or pores of other simplified geometry, with a distribution of cross-sectional sizes (although more detailed models are used for crystalline nanoporous materials such as zeolites, see for example Boutin et al., this volume and [2]). A major step forward in the 50s and 60s, was the recognition of the importance micropores, especially by Sing and his co-workers. This in turn led to the standard classification of pore sizes into micropores (< 2 nm), mesopores (2–50 nm) and macropores (> 50 nm), the latter being beyond the size range easily detectable by the capillary condensation of sub-critical nitrogen. Notwithstanding the issues raised by the use of an idealized pore model, it is not always recognised that this classification has its origin specifically in the adsorption of nitrogen at liquid nitrogen temperature [3].

At this low temperature, nitrogen condenses in mesopores and isotherms for pores in this size range exhibit a hysteresis loop. Isotherm shapes and loop shapes vary considerably, and the mapping of these shapes onto underlying models, characterised by a minimal set of parameters, continues to be an intriguing problem. Most typically, the goals for a material with pores wholly in the mesopore size range would be a pore size distribution and pore network connectivity, where the latter is, to some extent, determined by the form of the hysteresis loop.

* http://academic.csuohio.edu/foa8/
** http://hermes.umcs.lublin.pl/~rudzinsk/3.htm
*** http://www.triprinceton.org/workshop2006/Default.htm
**** http://sepapuri.yonsei.ac.kr/pbast-1.htm

A simple program along these lines has been proposed by several authors in the past [4]. However, from what has been said above, it is clear that many difficulties stand in the way of its implementation and that alternative approaches may lead to more effective characterisation procedures.

Computer simulation has contributed significantly toward the recognition and resolution of fundamental problems in adsorption in porous materials. Some advances stemming from this source may be highlighted:

- the extraordinary robustness of the BET method of surface area measurement [5,6]
- the demonstration that pores at the lower end of the micropore range are filled by liquid nitrogen at extremely low pressure [7]
- the demonstration of the effect of confinement on melting transitions (see Śliwinska-Bartkowiak et al., this volume)
- the revelation that the Kelvin equation fails badly, even within the mesopore size range, and is really only valid for macropores [8]
- the analysis of isotherms using databases of simulated isotherms, following the earlier use (and validation) of methods based on databases generated by density functional theory [9,10,11,12]
- the recognition that using apparently nonporous (or low surface area carbons) as reference states for adsorbent-adsorbate interactions may be flawed and that the use of reference materials closer in surface structure to the adsorbents of interest can lead to much improved pore size distributions that can be predictive for some gases and gas mixtures [20,21]
- a significant contribution to the use of ambient temperature adsorption in the characterisation of microporous materials [13,14]
- the investigation of idealised models that have helped to throw light on the complexities of the adsorption hysteresis phenomenon [15]

Experimentally, adsorption techniques have not been restricted solely to nitrogen adsorption. Much interest has focussed on the adsorption of argon [16], again at low temperatures. The appeal of this adsorbate arises from its simple rare gas structure, and consequently that it should be easier to understand at the theoretical level than molecular dinitrogen. Similarly, both krypton and xenon adsorption have been explored in the hope that they would be more effective probes for micropores [1,17, see also Pellenq and Levitz, this volume]. More recently, advances in low temperature and pressure measuring techniques have led to pioneering efforts to establish helium adsorption as a tool to augment characterisation methods. [18]

The use of high pressure room temperature carbon dioxide adsorption has been recommended by a number of authors [3,19,14, see Samios et al., this volume]. Sweatman and Quirke [14, 20, 21], for example, have investigated the ability of pore size distributions (PSDs) obtained from one gas to predict the adsorption of the other gases at the same temperature. They found that carbon dioxide PSDs are the most robust in the sense that they can predict the adsorption of other small molecules such as methane and nitrogen with reasonable accuracy.

Although the majority of experimental measurements of adsorption concentrate on adsorption isotherms, heats of adsorption offer a supplementary source of information that is often neglected. One reason for this, of course, is that the isotherm route to heats of adsorption requires that at least three isotherms be measured at three different temperatures, which confronts the experimentalist with the problem of arranging thermostatting, as well as the time-consuming additional measurements. An alternative, extensively exploited by Rouquerol and co-workers in Marseille, is to make direct calorimetric measurements of adsorption heats [22]. It has been shown that micropore distributions based on heats exhibit characteristic signatures that are absent from adsorption isotherms [23].

A concept that has played a prominent role in discussions of adsorption over many decades is that of surface heterogeneity. In its most elementary form, the effect of heterogeneity is manifested by sites of different energy on the adsorbent surface. For example, steps and kinks on a disordered surface have more atom-atom contacts with an adsorbate an open site on an identical open surface, and therefore a lower adsorption energy. This kind of effect becomes much more dramatic when isolated ionic sites, such as may occur in some zeolite cages, interact with a polar molecule, as shown for example by comparison of simulations of nitrogen adsorption in pure silica- and Ca-chabazite [24]. At the same time it has to be recognised that structural heterogeneity effects of this type can become intermingled with geometrical effects. This is illustrated by what is often referred to as the fundamental equation of adsorption,

$$\Gamma(p) = \int \Gamma(p, \varepsilon) f(\varepsilon) d\varepsilon \qquad (1.1)$$

Here $\Gamma(p, \varepsilon)$ is a local isotherm for adsorption at a site of energy ε at pressure p, and $f(\varepsilon)$ is a distribution of site energies. However, in micropores especially, the adsorption energy varies with pore size, and an exactly analogous expression to (1.1) can be written in which ε is replaced by R, a measure of pore width. The problem is further obscured by adsorbate-adsorbate interactions, since the total energy from this source is lowered as adsorbate loading increases.

Much theoretical work has been carried out on the effects of heterogeneity on adsorption, particularly by Polish and Russian groups, and several equations, often incorporating model site energy distributions, have been offered as expressions for the overall adsorption. When put to the test by comparison with simulation data [25,26] these are generally found to be barely adequate for the task [27], and more robust models have been proposed [28].

A signature considered to be characteristic of surface heterogeneity is a steep decline in the heat of adsorption with coverage (see for example

Siperstein et al., this volume). Model studies suggest that atomic disorder alone is insufficient to give this, but that suitable distributions of micropores that include very narrow pore sizes can do so [29]. This in turn suggests that characteristic "heterogeneous" heat curves may result from adsorbate occupying inter-particle crevices between nanoscale particles.

Whilst simple models can be enlightening, they leave open the question of whether additional "emergent" phenomena may be occurring due to the effects of interactions between adsorbate held in the complex interstices of real materials. With the rapid increase in computing power that has become available in recent years, a number of studies, motivated by this consideration, have been made in which collective representation of porous materials has been sought. Models of this type include: disordered media of random spheres [30,31], simulations in which the laboratory preparation of materials is imitated in simulation [32,33] and reverse Monte Carlo and simulated annealing, in which the structure factors of the real material are used as target functions in the reconstruction of the simulated adsorbent [34]. Subsequent simulation of adsorption in these materials can be compared with simpler models and interpreted in terms of known correlations relating to the solid structures. [35,36].

1.2   Transport

A detailed understanding of molecular transport through nanostructured materials is fundamental to the rational design of new materials and devices for separation processes (see for example Travis and Gubbins this volume). It is also central to the new field of nanofluidics and applications to problems involving, for example, high throughput characterisation, analysis and sequencing. Our current understanding of the processes involved in fluid motion (where we include both steady state [37] and transient flow) in nanoporous solids, in particular, the appropriate boundary conditions to use with hydrodynamic models, and the kinetics of fluid imbibition has progressed through advances in both theory [38,39,40,41] and simulation [42,43,44,45,46]. However, the precise relation between the transport of fluids in nanopores and the details of the interactions of the fluid with the pore wall remains an open problem. A key parameter characterising the applicability of the continuum equations (Navier Stokes equation [47]) is the Knudsen number Kn, defined as the ratio of the molecular free path to the transverse dimensions of the system. Computer simulation studies of Couette and Poiseuille flow in model slit pores [48,49,50] have shown that the flow of fluids in the small Kn regime is described remarkably well by macroscopic phenomenological hydrodynamics even for pore widths down to ten molecular diameters. What has remained unanswered by these theories, however, is the question of the precise relation between molecular properties of the interface and the hydrodynamic boundary conditions, inevitably appearing when the solid is approximated by a continuum. [51–54] The usual assumption made in continuum fluid dynamics is that the fluid velocity vanishes at the solid wall: the so-called "no-slip" boundary condition. In molecular models, where the solid is modelled as a continuum, the boundary conditions are usually specified by postulating a scattering law, *e.g.* diffuse or Knudsen's cosine law [55,56] boundary conditions, which may not be correct or even physically reasonable. Since it is clear that both equilibrium and non-equilibrium properties of the system can depend strongly on the imposed boundary conditions [43] it is vital that realistic boundary conditions are employed in molecular simulations.

An approach which leads to simple yet physically correct boundary conditions can be based on Maxwell's theory of slip. In the appendix to his paper on stresses in rarefied gases, published in 1879, [57] Maxwell developed a theory in which slip boundary conditions were obtained as a consequence of the molecular corrugation of the solid. This theory is based on the assumption that a particle, colliding with the wall, is thermalised by the wall with some probability α, or specularly reflected with probability $(1 - \alpha)$, resulting in a scattering law, which is a mixture of slip and stick boundary conditions. Since the derivation was made within the framework of the kinetic theory of gases, the theory does not of course, include the effects of adsorption

According to Maxwell's theory of slip, the distribution function for the velocity component parallel to the wall in the direction of flow is a linear combination of specularly reflected particles and those thermalised by the wall,

$$f_M(v_\parallel) = (1-\alpha)f(v_\parallel - u_{in}) + \alpha f(v_\parallel), \qquad (1.2)$$

where it is assumed that both distributions can be described by the Maxwellian,

$$f(w) = \sqrt{\frac{m}{2\pi kT}} \exp\left(-\frac{mw^2}{2kT}\right),$$

and u_{in} is the mean streaming velocity of the particles approaching the wall.

When the streaming velocity at the wall is significantly smaller than the mean thermal velocity, it is easy to show from Equation (1.2) and the definition of the mean velocity that the following relation holds

$$u_{out} = (1 - \alpha)\, u_{in}, \qquad (1.3)$$

where u_{out} is the mean tangential velocity of scattered molecules in the direction of flow. From Equation (1.3) we can find α as $\alpha = \Delta u/u_{in}$, where we used $\Delta u \equiv u_{in} - u_{out}$. The streaming velocity at the wall, u_w, is defined as a simple mean of u_{in} and u_{out}, and taking into account Equation (1.3), we obtain

$$u_w = (1 - 0.5\alpha)u_{in}. \qquad (1.4)$$

Table 1.1 Values of the Maxwell Coefficient for Nanopores

Material	Geometry	Width/ Diameter	Fluid	T	Density	α
		nm		K	kg/m^3	
[+]WC [62]	slit	200	Ne*	293	28	0.009
WC [62]	slit	200	Ar*	293	56	0.006
WC [62]	slit	200	Kr*	293	117	0.007
WC [62]	slit	200	N$_2$*	293	39	0.008
Graphite[‡]	slit	7.1	CH$_4$*	298	296	0.013
Rare gas [58][‡]	slit	7.1	CH$_4$*	298	296	0.54
SWCT(16,16)[◊]	cylinder	2.172	N$_2$[◊]	300	408	0.0016
SWCT(16,16)[◊]	cylinder	2.172	N$_2$[◊]	300	272	0.0015
SWCT(16,16)[◊]	cylinder	2.172	N$_2$[◊]	300	170	0.0017
Rare gas ('16,16')[◊]	cylinder	2.172	N$_2$[◊]	300	170	0.022
SWCT(16,16)[Δ]	cylinder	2.172	CH$_4$*	300	233	0.0023
SWCT(10,10)[Δ]	cylinder	1.36	CH$_4$*	300	207	0.0012
SWCT(7,7) [46]	cylinder	0.951	C$_{10}$H$_{22}$	300	106	0.000002
Rare gas ('7,7') [46]	cylinder	0.951	C$_{10}$H$_{22}$	300	618	0.0002

*modelled as a single LJ site, [+](0001) crystal basal plane of tungsten carbide, [‡]Table 2, reference 43, [◊]Table 2 reference 59 [Δ]Table 1 reference 61

One way of calculating α in a non equilibrium molecular dynamics simulation (NEMD) is to calculate velocities of sub-ensembles of molecules colliding with the wall and leaving it after the collision to obtain u_{in} and u_{out}. The friction force per particle exerted on a surface can be calculated from α and u_w using [43]

$$F_y = -\chi m \frac{2\alpha}{2-\alpha} u_w \qquad (1.5)$$

where χ is the collision frequency and m the mass of the colliding molecules. For N gas molecules in slit pores χ/N is of the order of 1 ps^{-1}.

Maxwell's coefficient α has been calculated using NEMD [43] and EMD [59,44] for a range of simple fluids as well as nanoparticle suspensions and solutions [60] flowing in (mostly) carbon nanopores (see table) and can be used to impose boundary conditions in a smooth wall model which reproduce the interfacial characteristics of the full molecular model [61]. However in order to get correct fluxes it is necessary to use values of α, which are almost twice those obtained from the full molecular simulation with molecular walls. This is likely to be the result of approximations in Maxwell's model including the fact that the wall friction is applied instantaneously to colliding particles.

From the table, the values of α for decane in carbon nanotubes are significantly smaller than for nitrogen and methane. This could be interpreted as being due to the larger size of the decane molecule with respect to the surface corrugation (D. Nicholson and S.K. Bhatia, "Scattering and

tangential momentum accommodation at a 2D adsorbate–solid interface,"
J. Membrane Sci. (in press)). In addition, the values of α for the graphitic pores
are significantly less than for the rare gas pores. This is because the surface
density of the rare gas pore is a fifth of that for the graphitic pore, leading to
a higher degree of corrugation of the surface potential and thus a higher value
of α. The value of α is a strong function of geometry since the corrugation of
the surface decreases as the curvature increases and this is evident in the
reduction of α between a 16,16 and a 10,10 nanotube for methane. Tungsten
carbide has a surface density, and hence corrugation, intermediate between
the rare gas pores and graphite ($\rho^s_{WC}/\rho^s_C = 0.36$). At the low densities reported
here, α for the WC slits is comparable to that for graphite slits.

The prediction of fluxes in nanoscale gaps has an important (and perhaps
unexpected) practical application to pressure standards where a knowledge
of the frictional drag force on a falling piston due to gas flowing in gaps, of
the order of 200 nm, is required [62]. Another application is to nanofluidics
where recent work has shown how a knowledge of α can be used to predict
the non equilibrium dynamics of capillary filling (imbibition) of nanotubes.
The radially averaged filling velocity is given by [46,63]

$$V = \frac{dL}{dt} = \frac{1}{2}\left[\frac{\frac{c}{a}(1-e^{-at})}{\left(\frac{c}{a}\right)^{1/2}\left(t+\frac{e^{-at}-1}{a}\right)^{1/2}}\right] \tag{1.6}$$

and the density profile by

$$\rho(x,t) = .5\rho ertc((x+x_0 - Vt)/\sqrt{4Dt}) \tag{1.7}$$

In the above equation $c = 4\Delta\gamma/\rho r$ $\Delta\gamma = \gamma_{sv} - \gamma_{sf}$ where γ_{sv} is the solid surface-
vapour surface tension and γ_{sf} is the surface tension for the imbibing fluid,
ρ is the mass density of the fluid, $a = \chi\alpha$, the number of wall collisions that
result in thermalization per molecule per second (χ is the mean number of
surface collisions per molecule per second, α, is Maxwell's coefficient[64]), V
is the average flow velocity of the fluid in the pore, x is the axial distance
along the pore from the wet pore entrance, x_0 the initial position of the
wetting fluid interface at $t = 0$, D is the diffusion coefficient (see reference
63 for a discussion of the meaning of this parameter in nanopores). Given a
$= \chi\alpha$ and c for an arbitrary nanotube, Equations (1.6,1.7) provide a complete
description of the wetting dynamics.

One challenge for the future is to extend this description of surface friction
to heterogeneous surfaces, especially at the nanoscale, and to more complex fluids
such as electrolytes (see Suh et al., this volume). We note that wetting by simple
fluids of striped and hexagonal chemically nanopatterned surfaces (homogeneous
solid fluid surface tensions γ_1, γ_2) has been shown [65] to obey Cassie's law [66]
for small surface contrast. Cassie's law states that the surface tension of a

patterned interface can be calculated from the properties of the homogeneous surfaces using the fractional coverage of the surface c by component one,

$$\gamma = c\gamma_1 + (1 - c)\,\gamma_2$$

There is some evidence from simulation [67] of a similar "Cassie law" behaviour for the Maxwell coefficient of a chemically patterned surface where the effective α is defined by

$$\alpha \sim c\alpha_1 + (1 - c)\alpha_2$$

However the general behaviour is likely to be very complex and requires detailed study. It is of particular importance to look at the effect of physical disorder and especially the role of surface features of nanoscale dimensions.

Another aspect of this problem is the breakdown of the hydrodynamic description for very small pores (<10 nm) where the velocity profiles can deviate significantly from parabolic [68,69] for pure fluids and mixtures (including colloidal mixtures) [70,60]. The choice of boundary condition then becomes more problematic since the boundary region may need to include these deviations. Elucidating the origins of these nanoscale changes (beyond remarking that the small pore has pronounced density oscillations) is an active area of research.

1.3   Summary

In section 1.1 we considered the role of simulation in allowing the pore size distribution of nanomaterials to be determined by inverting adsorption isotherms, and in section 2.1 we have concentrated on the prediction of surface friction and by implication, boundary conditions for fluid flow that can be used in analytic descriptions (as well as fluid dynamics simulations) of micro and nanofluidics systems. We have seen that molecular simulation has a unique role to play in the study of nanomaterials as it allows the calculation of physical properties which can be fed into more approximate approaches so important for the design of materials for engineering applications. In addition, the appropriate use of molecular simulation techniques produces a much enhanced physical understanding of the molecular processes underlying macroscopic phenomena and in doing so makes an invaluable contribution to progress in physical chemistry.

References

1. Rouquerol, F., Rouquerol, J. and Sing, K.S.W. (1998) *Adsorption by Powders and Porous Solids*, Elsevier, Amsterdam.
2. A. Fuchs, (2004). *Molecular Simulation of Zeolites* volume 30 nos. 9, 10. A special issue of *Molecular Simulation* devoted to zeolites.
3. Scaife, S., Kluson, P. and Quirke, N. (1999). "Characterisation of porous materials by gas adsorption: Do different molecular pores give different pore structures?," *J. Phys Chem.* B 2000, 313.

4. See for example Lopez-Ramon, M.V., Jagiello, J., Bandosz, T.J., and Seaton, N. A. (1997). "Determination of the pore size distribution and network connectivity in microporous solids by adsorption measurements and Monte Carlo simulation," Langmuir, 13, 4435. Nicholson, D. (1968). "Capillary models for porous media. II. Sorption desorption hysteresis in three-dimensional networks." *Trans. Far. Soc.* 64, 3416.

5. Rowley, L.A., Nicholson, D. and Parsonage, N.G. (1976) "Grand Ensemble Monte Carlo studies of physical adsorption. II. Structure of the adsorbate. Critique of theories of multilayer adsorption for 12-6 argon on a plane homogeneous solid." *Mol. Phy.* 31 389.

6. Gelb, L.D. and Gubbins, K.E. (1998) "Characterisation of porous glasses: Simulation models, adsorption isotherms, and the BET analysis method." *Langmuir* 14, 2097.

7. Nicholson, D. (1994). "A simulation study of nitrogen adsorbed in parallel sided micropores with corrugated potential functions" *J. Chem. Soc. Faraday Trans.*, 89, 181.

8. Peterson, B.K. and Gubbins, K.E. (1987) "Phase transitions in a cylindrical pore. Grand canonical Monte Carlo, mean field theory and the Kelvin equation." *Mol. Phys.* 62, 215.

9. Seaton, N.A., Walton, J.P.R.B., and Quirke, N. (1989) "A new analysis method for the determination of the pore-size distribution of porous carbons from nitrogen adsorption measurements," *Carbon* 27, 853.

10. Lastoskie, M.L., Gubbins, K.E., and Quirke, N. (1993) "Pore-size distribution analysis of microporous carbons — a density functional theory approach," *J. Phys. Chem.* 97, 4786.

11. Quirke, N. and Tennison, S.R.R. (1996) "The interpretation of pore-size distributions of microporous carbons," *Carbon* 34, 1281.

12. Lastoskie, M.L., Quirke, N., and Gubbins, K.E. (1997) "Structure of porous adsorbents: Analysis using density functional theory and molecular simulation," *Stud. Surf. Sci. Catal.* 104, 745.

13. Samios, S., Stubos, A.K., Kanellopoulos, N.K., Cracknell, R.F., Papadopoulos, G. K. and Nicholson, D. (1997) "Determination of micropore size distribution from grand canonical Monte Carlo simulations and experimental CO_2 isotherm data." *Langmuir* 13, 2795.

14. Sweatman, M.B. and Quirke, N. (2000) "Characterization of porous materials at ambient temperatures and high pressure," *J. Phys. Chem.* B. 105, 1403.

15. Sarkisov L. and Monson P.A. (2000) "Capillary condensation and hysteresis in disordered porous materials." *Studies in Surf Sci. and Catalysis* 128 21;. Sarkisov L. and Monson P.A. (2000) "Hysteresis in Monte Carlo and molecular dynamics simulations of adsorption in porous materials" *Langmuir* 16 9857.

16. Grillet, Y., Rouquerol, F. and Rouquerol, J. (1979) "Two dimensional freezing of nitrogen or argon on differently graphitized carbons." *J. Colloid and Interf. Sci.* 7., 239.

17. Olivier, J.P. (1999) "Thermodynamic properties of confined fluids I: Experimental measurements of krypton adsorbed by mesoporous silica from 80K to 130K." *World Sci.* 1.

18. Kuwabara, H., Suzuki, T., and Kaneko, K. (1991) "Ultramicropores in Microporous Carbon-Fibers Evidenced by Helium Adsorption at 4.2 K," *J. Chem. Soc. Faraday Trans.*, 87, 1915. Setoyama, N., Naneko, K., and RodriguezReinoso, F. (1996) "Ultramicropore characterization of microporous carbons by low-temperature helium adsorption," *J. Phy. Chem.*, 100, 10331. Kaneko, K. (2000)

"Specific intermolecular structures of gases confined in carbon nanospace," *Carbon*, 38, 287.

19. Cazorla-Amoros, D., Alcaniz-Monge, J., Linares-Solano, A. (1996),' Characterisation of activated carbon fibres by CO_2 adsorption' *Langmuir*, 12, 2820. Cazorla-Amoros D., Alcaniz-Monge, J., de la Casa-Lillo, M. A., Linares-Solano, A (1998),.' CO_2 as an adsorptive to characterise carbon molecular sieves and activated carbons' *Langmuir* 14, 4589; Garcia-Martinez, D., Cazorla-Amoros, D.; Linares-Solano, 2000 A. *Stud. Surf. Sci. Catal.*, 128, 485.

20. Sweatman, M.B. and Quirke, N. (2005) Gas adsorption in active carbons and the slit-pore model 1: Pure gas adsorption, *J. Phys. Chem.* 109, 10381.

21. Sweatman, M.B. and Quirke, N. in *Handbook of Theoretical and Computational Nanotechnology*, American Scientific Publishers 2005.

22. Tosi-pellenq, N., Grillet, Y., Rouquerol, J. and Llewellyn, P. (1992) "A microcalorimetric comparison of the adsorption of various gases on two microporous adsorbents; a model aluminosilicate and a natural clay." *Thermochim. Acta* 204, 79; Llewellyn, P., Coulomb, J.P., Grillet, Y., Patarin, J., Lauter, H. J., Reichart, H. and Rouquerol, J. (1993) "Adsorption by MFI-type zeolites examined by isothermal microcalorimetry and neutron diffraction. 1. Ar, Kr and methane." *Langmuir* 9, 1846–1851.

23. Nicholson, D. and Quirke, N. (2000) "The role of enthalpy of adsorption in micropore characterisation: A simulation study." *Studies in Surf Sci. and Catalysis* 128, 11.

24. T. Grey, J., G. Gale and D. Nicholson (2002) "Simulation studies of nitrogen in zeolites: Comparison of ionic and pure silica forms." *Fundamentals of Adsorption* 7, Eds. K. Kaneko, H. Kanoh, Y. Yazawa, IK International, 360.

25. Kruk, M., Jaroniec, M. and Choma, J. (1997) "Critical discussion of simple adsorption methods used to evaluate the micropore size distribution." *Adsorption* 3, 209.

26. Hutson, N.D. and Yang, R.T. (1997) "Theoretical basis for the Dubinin-Radushkevitch (D-R) adsorption isotherm equation." *Adsorption* 3, 189.

27. Bojan, M.J., Vernov, A.V. and Steele, W.A. (1992) "Simulation of adsorption in rough-walled cylindrical pores," *Langmuir* 8, 901.

28. Steele, W.A. (1999) "The supersite approach to adsorption on heterogeneous surfaces," *Langmuir* 15, 6083.

29. Nicholson, D. (1999) "A simulation study of energetic and structural heterogeneity in slit-shaped pores" *Langmuir*, 15, 2508.

30. Macelroy, J.M.D. and Raghavan, K. (1990) "Adsorption and diffusion of a Lennard-Jones vapour in microporous silica" *J. Chem. Phys.* 93, 2068.

31. Vuong, T. and Monson, P. A. (1998) "Monte Carlo simulations of adsorbed solutions in heterogeneous porous materials." *Adsorption* 16, 4880; Vuong, T. and Monson, P.A. (1999) "Monte Carlo simulations of adsorbed solutions in heterogeneous porous materials." *Adsorption* 5, 295.

32. Gelb, L.D. and Gubbins, K.E. (1998) "Characterisation of porous glasses: Simulation models, adsorption isotherms, and the BET analysis method." *Langmuir* 14, 2097–2111; Gelb, L.D. and Gubbins, K.E. (2002) "Molecular simulation of capillary phenomena in controlled pore glasses." *Fundamentals of Adsorption* 7, 333.

33. Pellenq, R.J.-M. and Levitz, P. (2001) "Adsorption/condensation of xenon in a disordered silica glass having a mixed (micro and meso) porosity." *Mol. Sim.* 27, 353.

34. Pikunic, J., Pellenq, R.J.-M., Thomson, K.T., Rouzaud, J.-N., Levitz, P. and Gubbins, K.E. (2001) "Improvoed molecular models for porous carbons." *Studies in Surf Sci. and Catalysis* 132, 647; Pikunic, J., Pellenq, R.J.-M. and Gubbins, K.E. (2002) "Modelling porous carbons by combining reverse Monte Carlo and simulated annealing." *Fundamentals of Adsorption* 7, 377. Gavalda, S., Gubbins, K.E., Hanzawa, Y., Kaneko, K., and Thomson, K.T. (2002) "Nitrogen adsorption in carbon aerogels: A molecular simulation study," *Langmuir*, 18, 2141. Brennan, J.K., Thomson, K.T., and Gubbins, K.E. (2002) "Adsorption of water in activated carbons: Effects of pore blocking and connectivity," *Langmuir*, 18, 5438. Pikunic, J., Gubbins, K.E., Pellenq, R.J.-M., Cohaut, N., Rannou, I., Gueth, J.M., Clinard, C., and Rouzaud, J.N. (2002) "Realistic molecular models for saccharose-based carbons," *App. Surf. Sci.*, 196, 98. Peterson, T., Yarovsky, I., McCulloch, D. G., and Opletal, G. (2003) "Structural analysis of carbonaceous solids using an adapted reverse Monte Carlo algorithm," *Carbon*, 41, 2403. Pikunic, J., Clinard, C., Cohaut, N., Gubbins, K.E., Guet, J.M., Pellenq, R.J.-M., Rannou, I., and Rouzaud, J.N. (2003) "Structural modeling of porous carbons: Constrained reverse Monte Carlo method," *Langmuir*, 19, 8565. Pikunic, J. and Gubbins, K.E. (2003) "Molecular dynamics simulations of simple fluids confined in realistic models of nanoporous carbons," *Eur. Phys. J. e*, 12, 35. Petersen, T., Yarovsky, I., Snook, I. K., McCulloch, D.G., and Opletal, G. (2004) "Microstructure of an industrial char by diffraction techniques and reverse Monte Carlo modeling," *Carbon*, 42, 2457.

35. Gelb, L.D. and Gubbins, K.E. (1999) "Pore size distribution in porous glasses: Simulation models a computer simulation study." *Langmuir* 15, 305.

36. Levitz, P. and Tchoubar, D. (1992) "Disordered porous solids: from chord distribution to small angle scattering," *J. Phys.* I 27, 771.

37. Evans, D.J. and Morriss, G.P. (1990) *Statistical Mechanics of Nonequilibrium Liquids*, Academic, London 1990.

38. Vollmer, J. (2002) "Chaos, spatial extension, transport, and non-equilibrium thermodynamics," *Phys. Rep.* 372, 131.

39. Kärger, J. and Ruthven, D.M. (1992) *Diffusion in Zeolites and Other Microporous Solids*, New York, Wiley, 2005.

40. Churaev, N.V. *Liquid and Vapour Flows in Porous Bodies*, Gordon and Breach, Amsterdam 2000.

41. Kornev, K.G. and Neimark, A.V. (2003), "Modeling of spontaneous penetration of viscoelastic fluids and biofluids into capillaries," *J. Colloid Interface Sci.*" 262, 253.

42. Thompson, P.A. and Troian, S.M. (1997) "A general boundary condition for liquid flow at solid surfaces," *Nature* 389, 360.

43. Sokhan, V.P., Nicholson, D. and Quirke, N. (2001) "Fluid flow in nanopores: an examination of hydrodynamic boundary conditions," *J. Chem. Phys.* 115, 3878.

44. Sokhan, V.P. and Quirke, N. (2004), "Interfacial Friction and Collective Diffusion in Nanopores," *Mol Sim*, 30, 217.

45 Supple, S. and Quirke, N. (2003) "Rapid inhibition of fluids in carbon nanotubes," *Phys. Rev. Lett.* 90, 214501.

46. Supple, S. and Quirke, N. (2004), "Nanocapillarity: Fluid imbibition in single wall nanotubes 1: Imbibition speeds for single wall carbon nanotubes," *J Chem Phys*, 121, 8571.

47 KM White. *Fluid Mechanics* 5th edition McGraw Hill, 2003.

48. P.A. Thompson and S.M. Troian, (1997). "A general boundary condition for liquid flow at solid surfaces," *Nature* 389, 360.
49. B.D. Todd and D.J. Evans, (1995). "The heat flux vector for highly inhomogeneous nonequilibrium fluids in very narrow pores," *J. Chem. Phys.* 103, 9804.
50. K.P. Travis, B.D. Todd, and D.J. Evans, (1997). "Departure from Navier-Stokes hydrodynamics in confined liquids," *Phys. Rev. E* 55, 4288.
51. U. Heinbuch and J. Fischer, (1989). "Liquid flow in pores: Slip, no-slip, or multilayer stickling," *Phys. Rev* A 40, 1144.
52. P.A. Thompson, M.O. Robbins, (1990). "Shear flow near solids: Epitaxial order and flow boundary conditions," *Phys. Rev.* A 41, 6830.
53. L. Bocquet and J.-L. Barrat, (1993). "Hydrodynamic boundary conditions and correlation functions of confined fluids," *Phys. Rev. Lett.* 70, 2726; (1994). "Hydrodynamic boundary conditions, correlation functions, and Kubo relations for confined fluids," *Phys. Rev* E 49, 3079.
54. J.-L. Barrat and L. Bocquet, (1999). "Large Slip Effect at a Nonwetting Fluid-Solid Interface," *Phys. Rev. Lett.* 82, 4671; (1999) *Faraday Disc.* 112.
55. S.-H. Suh and J.M.D. MacElroy, (1986). "Molecular dynamics simulation of hindered diffusion in microcapillaries," *Mol. Phys.* 58, 445; (1987). "Computer simulation of moderately dense hard-sphere fluids and mixtures in microcapillaries," *Mol. Phys.* 60, 475.
56. J.P. Valleau, D.J. Diestler, J.H. Cushman, M. Schoen, A.W. Hertzner, and M.E. Riley, (1991). "Comment on: Adsorption and diffusion at rough surfaces. A comparison of statistical mechanics, molecular dynamics, and kinetic theory," *J. Chem. Phys.* 95, 6194; J.H. Thurtell and G.W. Thurtell, (1988). "Adsorption and diffusion at rough surfaces: A comparison of statistical mechanics, molecular dynamics, and kinetic theory," *J. Chem. Phys.* 88, 6641.
57. J.C. Maxwell. *Phil. Trans. Roy. Soc.*, (1879). Reprinted in: *The Scientific Papers of James Clerk Maxwell* (Cambridge University Press, 1890) 2, 703.
58. To compare with nongraphitic tubes that would present a more corrugated potential energy surface, "rare gas tubes" are used, so-called because the *surface* that is rolled to form the tube has a low surface density of atoms, indicative of a rare gas wall, see reference 59.
59. V.P. Sokhan, D. Nicholson and N. Quirke, (2004). "Transport properties of nitrogen in single walled nanotubes" *J. Chem. Phys.* 120, 3855.
60. T. Myoshi and N Quirke (unpublished work).
61. V.P. Sokhan, D. Nicholson and N. Quirke, 2002. "Fluid flow in nanopores: Accurate boundary conditions for carbon nanotubes," *J. Chem. Phys.* 117, 8531,
62. V.P. Sokhan, N. Quirke and J. Greenwood, "Viscous drag forces in gas operated pressure balances," *Mol. Sim.*, (31,535). "Values of α from V.P. Sokhan," N. Quirke and J. Greenwood, (to be published).
63. Supple, S. and N. Quirke (2005), "Nanocapillarity: II: Density profile and molecular structure for decane in carbon nanotubes," *J. Chem. Phys.* 122, 104706.
64. Sokhan, V.P. Nicholson, D. and Quirke, N. etc. (2001). "Fluid flow in nanopores: An examination of hydrodynamic boundary conditions," *J. Chem. Phys.* 115, 3878.
65. M. Schneemilch and N. Quirke, 2003. "The Interaction of Fluids with Nanomaterials: Contact Angles at Nanopatterned Interfaces," *Mol. Sim.*, 29, 685–695.
66. Cassie, A.B.D. (1948), Contact angles, *Faraday Discuss. Soc.* 3, 11.
67. Norton, J. and Quirke, N. (unpublished work).

68. Zhang, J.F., Todd, B.D., Travis, K.P. "Viscosity of confined inhomogeneous nonequilibrium fluids," *J. Chem. Phys.* 121, 10778.
69. See figure 2, reference 44, for the distorted flow profiles in a slit pore $H = 4$ nm.
70. Kairn, T. Daivis, P.J. McPhie, M. Snook, I.K. "Poiseuille Flow of Colloidal Fluids in Micro-channels," Poster: Pacific Rim conference on Nanoscience, Broome 2004.

chapter two

Modelling gas adsorption in slit-pores using Monte Carlo simulation

M.B. Sweatman

N. Quirke*

Imperial College

Contents

2.1   Introduction

In this chapter we discuss the use of Monte Carlo simulation to model the equilibrium adsorption of gases in slit pores. While there are no perfect slit pore systems in nature, the ideal slit pore model is a useful approximation to pores in real adsorbents of practical interest such as activated carbons.

* Corresponding author.
Reprint from *Molecular Simulation*, 27: 5–6, 2001. http://www.tandf.co.uk

Over the last twenty-five years the understanding of equilibrium fluid behaviour in restricted geometries has advanced considerably. Early theoretical models [1,2], such as the Langmuir and Brunquer-Emmett Teller (BET) isotherms and the Kelvin equation, have been superseded by modern approaches such as density functional theory [3] (DFT) and Monte Carlo simulation [4–6]. Indeed, one of the early applications of the Monte Carlo method to the study of physical adsorption on graphite by Rowley, Nicholson and Parsonage [7] in 1976 evaluated multi-layer methods, including the BET, Dubinin and Frenkel-Halsey-Hill (FHH) models, by comparing their predicted adsorption isotherms and heats of adsorption with simulation data for Lennard– Jones argon. None were considered to be satisfactory. In the present article we present an overview of current modelling procedures used to predict the adsorption of nitrogen (N_2), carbon-monoxide (CO), methane (CH_4) and carbon-dioxide (CO_2) in graphitic pores.

The Metropolis Monte Carlo technique [8] originated with Metropolis et al. in 1953 and was extended to the Grand-canonical ensemble by Norman and Filinov [9], Rowley et al. [10] and Adams [11]. Application to adsorption problems soon followed [7]. The Grand-canonical ensemble is the natural ensemble with which to study adsorption in open slit pores because the ensemble is specified by chemical potential, volume and temperature. In a slit pore adsorbate pressure is generally not equal to reservoir pressure and, unless experiments are performed with surface force apparatus [12], the natural comparison of simulation and experiment is made through the chemical potential. Gases absorbed on surfaces have a non-uniform density profile in the direction normal to the surface. In order to properly describe this inhomogeneity, molecular models must be accurate for fluid densities varying from gas to dense liquid. In developing our model potentials for gas adsorption we require that they are capable of predicting bulk fluid phase coexistence properties. In this way we ensure that both vapour and liquid-like regions of absorbed fluids are accurately described. The appropriate Monte Carlo technique for predicting bulk phase coexistence properties is the Gibbs ensemble method [13–15] invented by Panagiotopoulos in 1987 (and also applied to non-uniform fluids [16,17]).

This work describes molecular models for adsorption of N_2, CO, CH_4 and CO_2 in graphitic slit-pores. A great deal of work has been performed on these or similar systems [18–20] (as well as rare gases [21,22]). Our focus is on the determination of accurate two-body effective potentials calibrated against experimental data for bulk phase coexistence properties and adsorption on standard graphitic surfaces. We describe our modelling methodology in the next section together with an overview of the Gibbs and Grand-canonical ensemble simulation techniques. These techniques are used to fine-tune our effective molecular models. In the final section we describe our results for gas adsorption in graphitic slit pores for a range of pore widths and bulk pressures and comment on their implications for the characterisation of porous materials using gas adsorption isotherms.

2.2   Methods

2.2.1   Molecular models

We wish to determine useful molecular models for the adsorbates N_2, CO, CH_4 and CO_2. We model all repulsive–dispersive interactions with the Lennard–Jones (LJ) potential,

$$\varphi_{ij}^{LJ}(r_{ij}) = 4\varepsilon_{ij}((\sigma_{ij}/r_{ij})^{12} - (\sigma_{ij}/r_{ij})^6) \qquad (2.1)$$

where σ_{ij} and ε_{ij} define the length and energy scale respectively of the interaction between LJ sites i and j, separated by r_{ij} on different molecules, and all electrostatic interactions with partial charges,

$$\varphi_{ij}^{C}(r_{ij}) = C_i C_j / 4\pi\varepsilon_0 r_{ij} \qquad (2.2)$$

where i and j are charge sites (not necessarily coincident with any LJ sites), with charge C, on different molecules and ε_0 is the vacuum permittivity. We constrain cross-interactions between unlike LJ sites to be related to the pure interaction parameters by the Lorentz–Bethelot rules

$$\sigma_{ij} = (\sigma_{ii} + \sigma_{jj})/2; \qquad \varepsilon_{ij} = \sqrt{\varepsilon_{ii}\varepsilon_{jj}} \qquad (2.3)$$

Adsorbate molecular models are constructed from at least one LJ site, with the position, $(\vec{r}_{0i})$, of each site fixed relative to the centre-of-mass and orientation of the molecule. The interaction between two (different) molecules, α and β, is then simply the sum of the individual LJ and charge pair-interactions,

$$\phi_{\alpha\beta} = \sum_{i_\alpha j_\beta} \phi_{ij}^{LJ} + \phi_{ij}^{C} \qquad (2.4)$$

where i_α indicates site i on molecule α

The total interaction energy between adsorbate molecules is then the sum

$$U_{gg} = U_{SR} + U_{LR}^{LJ} + U_{LR}^{C} = \sum_{\alpha<\beta} \phi_{\alpha\beta} \qquad (2.5)$$

The interaction between a pair of molecules must be subject to a cut-off, r_c. When the distance between the centre-of-mass of each molecule is less than r_c, pair-interactions are calculated explicitly according to Equation 2.4 and summed to give U_{SR}. However, for molecules outside this range LJ pair-interactions are treated in an average sense to give a long-range contribution, U_{LR}^{LJ}, to U_{gg}. Setting the pair correlation function between molecules separated by more than r_c, to unity gives, for a bulk fluid,

$$U_{LR}^{LJ} = 2\pi \sum_{\alpha}\sum_{i_\alpha}\sum_{j} \rho_j \int_{r_c}^{\infty} dr\, r^2 \phi_{ij}^{LJ}(r) = \frac{8\pi}{3}\sum_{i} N_i \sum_{j} \rho_j \varepsilon_{ij}\left(\frac{\sigma_{ij}^{12}}{3r_c^9} - \frac{\sigma_{ij}^6}{r_c^3}\right) \quad (2.6)$$

where N_i is the number of sites of type i in the simulation and ρ_j is the density of LJ site type j in the corresponding bulk phase. The bulk or long-range density is required in advance and can be obtained from earlier simulations.

In this work we neglect the long-range contribution of electrostatic pair-interactions since it is always small relative to other contributions. For example, in liquid CO_2 at 265 K the contribution of electrostatic interactions is about 20% of the total interaction energy, and long-range LJ interactions contribute about 1% of the total interaction energy (using the cut-off defined in the Results section). However, for systems where long-range electrostatic interactions are thought to be significant (for example in water) methods such as the Ewald summation method [23–25] or the reaction field method [26] can be employed. When the distribution of partial charges of a molecule are quadrupolar, the long-range contribution can be determined in a similar fashion to the long-range LJ contribution (6), i.e., the long-range electrostatic interaction between two quadrupolar molecules can be approximated by the pair interaction of two quadrupolar moments and long-range pair correlation functions can be set to unity.

It is not uncommon to neglect long-range interactions altogether, leaving a small step discontinuity in the interaction between two molecules. To reduce the unwelcome effect of such discontinuities, the remaining short-range part of the potential can be shifted. Unless the range of each LJ and electrostatic pair-interaction is calculated individually to determine its short or long-range nature, rather than according to the separation of molecular centre-of-masses, very small discontinuities in the pair-interaction of two molecules will persist.

For the simple molecular gases that are the focus of this work we are free to choose interaction parameters σ_{ii}; ε_{ii}; C_i, $\vec{r}_{0i}$. We tune the interaction parameters for each adsorbate so that bulk liquid–gas coexistence properties fit experimental data, using Gibbs ensemble simulation to determine liquid–gas coexistence properties of a given model. We will show that with these molecular models we can reproduce bulk fluid experimental coexistence data for each adsorbate with reasonable accuracy. Clearly, for more demanding problems, more accurate interaction potential is required. For example, simulations of hydrocarbons often employ stretch and torsional potentials energy terms, while those of water sometimes employ hydrogen-bonding and polarization terms.

We model the interaction between each molecular LJ site (i) and each graphite surface (s) by a Steele potential [27]

$$V_{i_\alpha s}\left(z_{i_\alpha}\right) = 2\pi\rho_c \Delta \varepsilon_{is} \sigma_{is}^2 \left(\frac{2}{5}\left(\frac{\sigma_{is}}{z_{i_\alpha}}\right)^{10} - \left(\frac{\sigma_{is}}{z_{i_\alpha}}\right)^4 - \frac{\sigma_{is}^4}{3\Delta\left(z_{i_\alpha}+0.61\Delta\right)^3} \right) \quad (2.7)$$

where, for graphite, we set $\rho_c = 114$ nm^{-3} and $\Delta = 0.335$ nm and z_{i_α} is the distance of LJ-site i on molecule α from the plane of carbon atom centers in the first layer of the surface. The Steele potential is invariant in directions parallel to the surface and generally provides a good approximation to the potential obtained by summing individual LJ-surface atom pair interactions. The summed LJ potential can be very smooth in directions parallel to the slit, for example the Boltzmann factor for the methane-summed LJ interaction varies by at most 1% across the surface at the potential minimum at 298 K.

The ideal slit-pore potential is given by

$$V_{i_\alpha}^{\text{ext}}\left(w, z_{i_\alpha}\right) = V_{i_\alpha s}\left(z_{i_\alpha}\right) + V_{i_\alpha s}\left(w - z_{i_\alpha}\right) \quad (2.8)$$

where w is the width between carbon atom centres in the first layer of opposing parallel surfaces. This gives

$$U = U_{\text{gg}} + V_{\text{g}} = U_{\text{gg}} + \sum_\alpha \sum_{i_\alpha} V_{i_\alpha}^{\text{ext}}\left(w, z_{i_\alpha}\right) \quad (2.9)$$

for the total interaction energy of a given microscopic state. This is the potential-energy function that we investigate with Monte Carlo simulation. For fluid in a slit pore the long-range LJ contribution to U_{gg} is

$$\frac{1}{2}\left\langle \sum_\alpha \sum_{i_\alpha} \sum_j \int_{r_{ij}>r_c}^{\infty} d\vec{r}_j \rho_j(\vec{r})g_{ij}^{(2)}(\vec{r}_i,\vec{r}_j)\phi_{ij}^{\text{LJ}}(r_{ij}) \right\rangle \quad (2.10)$$

where $\rho_j(\vec{r})$ is the singlet density [28] of site j, $g_{ij}^{(2)}(\vec{r}_i,\vec{r}_j)$ is the pair distribution function between sites i and j at $\vec{r}_i$ and $\vec{r}_j$, $r_{ij} = |\vec{r}_i - \vec{r}_j| = |\vec{r}_{ij}|$ and the angle brackets denote an ensemble average. By setting correlations between pairs of molecules to unity if they are separated by more than r_c we obtain from Equation 2.10

$$U_{\text{LR}}^{\text{LJ}} = \frac{1}{2}\left\langle \sum_\alpha \sum_{i_\alpha} \sum_j \int_{-\infty}^{\infty} dz_j \rho_j(z_j) \int_{\max(|z_{ij}|,r_c)}^{\infty} dr 2\pi r \phi_{ij}^{\text{LJ}}(r) \right\rangle \quad (2.11)$$

The right-hand integral can be computed in advance for a range of values of the lower limit and $\rho_j(z)$ can be found from an earlier simulation. If $w < r_c$ or if $\rho_j(z)$ is not too inhomogeneous, then the molecular density further than r_c from site i_α can be "smeared" to give a uniform density, ρ_j, within the slit. Then Equation (2.11) becomes the sum of Equation (2.6) and two further contributions, each calculated as

$$-4\pi\left\langle \sum_\alpha \sum_{i_\alpha} \sum_j \rho_j \varepsilon_{ij} \left(\frac{\sigma_{ij}^{12}}{90z_1^9} - \frac{\sigma_{ij}^6}{12z_1^3} + (r_c - z_2)\left(\frac{\sigma_{ij}^{12}}{10r_c^{10}} - \frac{\sigma_{ij}^6}{4r_c^4} \right) \right) \right\rangle; \tag{2.12}$$

$$z_1 = \max(r_c, z), z_2 = \min(r_c, z),$$

where $z = z_{i_\alpha}$ and $z = w - z_{i_\alpha}$ for the separate contributions. Calculation of this expression is much faster than Equation (2.11). The same technique can be used to evaluate the long-range contribution to U_{gg} from quadrupolar pair interactions confined to a slit. Specialised techniques have been invented for treating confined fluids with more general electrostatic interactions [29].

Just as with fluid–fluid interaction parameters, the LJ site-surface inter-action parameters, $(\sigma_{is}, \varepsilon_{is})$ must be tuned so that agreement between simulation, in this case using the Grand-canonical ensemble, and reference data is reasonable. In this work we choose gas-surface interaction parameters [30,31] calibrated to experimental adsorption isotherms of the gases on a low surface area porous carbon, Vulcan 3G. So, our surface model is intended to represent the surface of porous carbons rather than graphite. Of course, such amorphous materials cannot be characterised by a single slit pore with fitted gas-surface interaction parameters, but previous work* has shown that characterizing such materials in terms of poly-disperse arrays of slit-pores is reasonably successful. More demanding applications require more accurate surface models. For example, the vibrational modes of carbon nanotubes *in vacuo* have been simulated using a Tersoff–Brenner potential [32], which approximates the many-body and co-ordinated nature of carbon interactions in molecular carbon and hydrocarbon materials.

2.3   The Gibbs ensemble

A Gibbs ensemble simulation simulates the coexistence of two bulk phases without simulating the interface between them. Each phase is simulated in a separate "box" with periodic boundary conditions and does not interact with the phase in the other box. Coexistence is guaranteed by the choice of Monte Carlo moves that produce the conditions for phase coexistence; equality of temperature, pressure and chemical potential. Intra-box moves (a molecule is moved randomly within the same box) equilibrate tempera-ture, inter-box moves (a molecule is moved to a random location in the other box) equilibrate chemical potential and volume moves (volume is transferred from one box to the other) equilibrate pressure (see Figure 2.1).

---

* For example, see Ref.[30] and references therein.

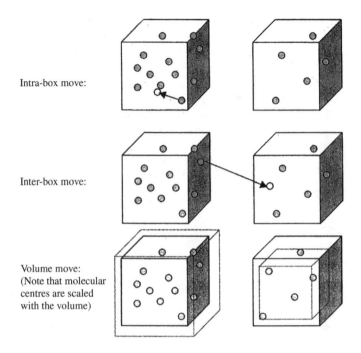

Intra-box move:

Inter-box move:

Volume move:
(Note that molecular
centres are scaled
with the volume)

Figure 2.1 Schematic representation of Gibbs ensemble moves.

We performed Gibbs ensemble simulations for the pure fluids N_2, CO, CH_4 and CO_2 over a range of sub-critical temperatures. Our aim is to achieve good agreement between the results of Gibbs ensemble simulations and experimental data [33–40], namely, the liquid and gas densities and pressure at coexistence, by adjustment of the effective interaction parameters (σ_{ii}, ε_{ii}, C_i, $\vec{r}_{0i}$) for each adsorbate. We employ long-range corrections for LJ interactions only. For each simulation we started the two simulation boxes with lattice-like configurations, the density of one box being 50% greater than the other box. The dimensions of each box are chosen so that their volumes are approximately equal at equilibrium and to ensure that the box length is never less than the cut-off radius. We choose different moves at random with predefined probability. For a simulation with N molecules the relative probabilities for intra-box, inter-box and volume moves are 1, a and $1/N$, respectively. The value of a is fixed so that the number of accepted inter-box moves is generally not less than 10% of the number of accepted intra-box moves. Thus, a is typically in the range 1–100, with higher values for denser liquid phases. Intra-box and inter-box moves are performed by choosing a molecule at random from both boxes. An intra-box move is performed by displacing a molecule in a random direction by a distance chosen randomly from a predefined interval. The molecule is then rigidly rotated about a randomly chosen Cartesian axis by an angle chosen randomly from a predefined interval. An inter-box move is performed by destroying a molecule in one box and then creating it at a random position with random orientation in the other box. A volume move is performed by taking a step on $\ln(V_{b1}/V_{b2})$ with step-length chosen at random from a predefined interval (the subscripts b1 and b2 refer to box 1 and box 2 respectively). When a box changes volume, the location of the centre of each molecule is scaled accordingly so that volume loss corresponds to compression and volume gain corresponds to decompression. The maximum step-size of intra-box and volume moves is fixed so that about 50% of these moves are accepted. We find that these move selection rules generally result in quick and stable phase separation.

Intra-box moves are accepted with probability

$$\min\{1, \exp[-\beta \Delta U]\} \tag{2.13}$$

where $\Delta U = U^n - U^m$ is the difference in total energy between the trial (n) state and the current (m) state. The total energy of a given state is the sum of the energies of the individual boxes. Inter-box moves are accepted with probability

$$\min\left\{1, \frac{N_{b1}V_{b2}}{(N_{b2}+1)V_{b1}} \exp[-\beta \Delta U]\right\} \tag{2.14}$$

if the molecule is to be moved from box 1 to box 2. The volume move acceptance rule must take account of the logarithmic volume-step selection rule and is

$$\min\left\{1, \left(\frac{V_{b1}^n}{V_{b1}^m}\right)^{N_{b1}+1} \left(\frac{V_{b2}^n}{V_{b2}^m}\right)^{N_{b2}+1} \exp[-\beta \Delta U]\right\} \tag{2.15}$$

These selection and acceptance rules satisfy microscopic reversibility and guarantee that the limiting distribution of states conforms to the Boltzmann distribution [4,5].

2.4   The Grand-canonical ensemble

A Grand-canonical ensemble simulation simulates a single phase at a given set of chemical potentials (one for each distinct species of molecule) $\{\mu_\alpha\}$, volume and temperature. Equilibrium is achieved by careful choice of intra-box moves (temperature equilibrium), and creation and annihilation moves (chemical potential equilibrium). Our aim is to generate databases of the adsorption of gases in a range of pore widths over a wide range of pressures. These databases can then be used, together with the poly-disperse slit-pore model, for the characterisation of porous carbons. We use Grand-canonical ensemble simulation to construct the databases.

Grand-canonical ensemble simulations are initialised with either an empty box or a configuration of molecules obtained from an earlier simulation. For simulations in a slit pore the width of the box is fixed. The remaining free dimensions of the box are chosen to ensure sufficient molecules within it at equilibrium and are never less than twice the cut-off range. We perform three different types of move at random with equal probability. An intra-box move is performed in an identical manner to an intra-box move in a Gibbs ensemble simulation. A creation move is performed by creating a molecule at a random position with random orientation in the box. An annihilation move is performed by choosing a molecule in the box at random and deleting it from the simulation. The maximum step-size of intra-box moves is fixed so that about 50% of these moves are accepted.

Intra-box moves are accepted according to Equation 2.13. Creation and annihilation moves are accepted with probability

$$\min\left\{1, \frac{V\Lambda_\alpha^{-3}}{N+1} \exp[\beta(\mu_\alpha - \Delta U)]\right\} \tag{2.16}$$

and

$$\min\left\{1, \frac{N+1}{V\Lambda_\alpha^{-3}} \exp[\beta(-\mu_\alpha - \Delta U)]\right\} \tag{2.17}$$

respectively. These selection and acceptance rules satisfy microscopic reversibility and guarantee that the limiting distribution of states conforms to the Boltzmann distribution [4,5]

2.5   Some thermodynamics

The grand potential, Ω, is related to the grand partition function, Ξ, via

$$\Omega = -k_B T \ln \Xi \tag{2.18}$$

which, with the Boltzmann distribution, gives

$$\Omega = F - \sum_\alpha \mu_\alpha \langle N_\alpha \rangle = \langle E \rangle - TS - \sum_\alpha \mu_\alpha \langle N_\alpha \rangle \tag{2.19}$$

where, F, E and S are the Helmholtz free energy, internal energy and entropy respectively and Ω is minimized at equilibrium [41]. In the thermodynamic limit, we can drop the ensemble average notation, $\langle \rangle$, and Ω acquires non-analyticities along the loci of phase transitions [41]. In this work we focus on the behaviour of fluid adsorbed in solid slit pores. The model potential (7) approximates the solid surface as an effective external potential. So, we construct Gibbs dividing surfaces [42,43] at $z = 0$ and $z = w$ and effectively ignore the contribution of the surface or reservoir to Equation 2.19. For a planar adsorbed system with area A and width w, an infinitesimal change in energy is written [43]

$$dE = T\,dS + \sum_{\alpha} \mu_{\alpha} dN_{\alpha} - \int_0^w P_T(z) dz dA - AP_N dw \qquad (2.20)$$

where P_T and P_N are the transverse (parallel) and normal components of the pressure tensor, respectively. The transverse component varies with z but the normal component does not. For fluid adsorbed on an isolated surface $P_N = P$ while for a bulk system all components are equal to P. With Equation 2.19 this gives

$$d\Omega = -S\,dT - \sum_{\alpha} N_{\alpha} d\mu_{\alpha} - \int_0^w P_T(z) dz dA - AP_N dw \qquad (2.21)$$

For fluid at given T and w we obtain two useful routes to the grand potential. The first by integrating along a continuous isotherm

$$d\Omega = -\sum_{\alpha} N_{\alpha} d\mu_{\alpha} \qquad (2.22)$$

and the second from P_T

$$\Omega = -A\int_0^w P_T(z) dz = -A(wP + \gamma) \qquad (2.23)$$

For a bulk system one has from Equation 2.21

$$dP = \sum_{\alpha} \rho_{\alpha} d\mu_{\alpha}; \qquad \gamma = 0 \qquad (2.24)$$

where γ is the negative of the surface tension

$$-\gamma = \left(\frac{\partial \Omega}{\partial A}\right)_{T,\mu_a,w} + wp$$

Equation (2.21) requires the non-analyticities that develop in Ω in the thermodynamic limit to be manifest in the behavior of S, N_{α}, P_T and P_N. At first-order phase transitions [44], S, N_{α} and P_N display step-discontinuities, while the average transverse pressure displays a step-discontinuity in its gradient with respect to T, μ_{α} and w. So, just as the bulk density is the order-parameter that signals the bulk liquid–gas phase transition, the average pore density, $\rho_{\alpha} = N_{\alpha}/Aw$, is the order parameter for phase transitions in a slitpore. Similarly, just as bulk pressure is maximised at equilibrium, the average transverse pressure is also maximised at equilibrium.

Because we reduce all gas–gas interactions to effective pair-potentials the components of the pressure tensor can be defined microscopically from the pair-virial [4,5]. The v-component for a planar geometry is (according to the "Irving and Kirkwood" definition [42])

$$p_v(z) = k_B T \sum_\alpha \rho_\alpha(z) - \frac{1}{2A} \left\langle \sum_{\alpha \neq \beta} \sum_{i\alpha j\beta} \left(\frac{d\phi_{ij}^{LJ}}{dr_{ij}} + \frac{d\phi_{ij}^C}{dr_{ij}} \right) \frac{v_{\alpha\beta} v_{ij}}{r_{ij} \, |z_{ij}|} \theta\left(\frac{z - z_i}{z_{ij}} \right) \theta\left(\frac{z_j - z}{z_{ij}} \right) \right\rangle$$

$$(2.25)$$

where $\rho_\alpha(z)$ is the singlet density of molecular species α, $v_{\alpha\beta}$ and v_{ij} are the v–components of the vectors between the centres of molecules α and β and the sites i and j respectively and θ is the Heaviside step-function. Using Equation 2.23 gives

$$\Omega = k_B T \sum_\alpha \langle N_\alpha \rangle - \left\langle \sum_{\alpha \neq \beta} \sum_{i\alpha j\beta} \left(\frac{d\phi_{ij}^{LJ}}{dr_{ij}} + \frac{d\phi_{ij}^C}{dr_{ij}} \right) \frac{x_{\alpha\beta} x_{ij} + y_{\alpha\beta} y_{ij}}{4r_{ij}} \right\rangle \qquad (2.26)$$

Of course, these expressions are useful for the short-range part of any pair-potential only. Long-range LJ corrections to Ω, obtained with the same approximations used in Equation 2.12, are two contributions calculated as

$$2\pi \left\langle \sum_{i\alpha} \sum_j \rho_j \varepsilon_{ij} \left(\frac{m^3}{3} \left(\frac{\sigma_{ij}^{12}}{r_c^{12}} - \frac{\sigma_{ij}^6}{r_c^6} \right) - 6m \left(\frac{\sigma_{ij}^{12}}{5r_c^{10}} - \frac{\sigma_{ij}^6}{4r_c^4} \right) \right. \right.$$

$$\left. \left. - \frac{1}{45} \left(\frac{\sigma_{ij}^{12}}{m^9} - \frac{\sigma_{ij}^{12}}{z^9} \right) + \frac{1}{6} \left(\frac{\sigma_{ij}^6}{m^3} - \frac{\sigma_{ij}^6}{z^3} \right) \right) \right\rangle$$

$$(2.27)$$

with $m = \min(z, r_c)$ and $z = z_{i\alpha}$ and $z = w - z_{i\alpha}$ for the two contributions. This expression can be calculated either for each configuration (and then averaged) or at the end of the simulation using the singlet density, $\rho_i(z)$. A similar expression can be obtained for the long-range contribution from quadrupole pair interactions. The contribution of more general electrostatic interactions to the components of the pressure tensor and the grand-potential can be found using the Ewald summation method [45] or other methods [29]. For a uniform fluid, all components of the pressure tensor are equal and the short-range contribution to P is

$$P_{SR} = k_B T \sum_i \langle \rho_i \rangle - \frac{1}{2V} \left\langle \sum_{\alpha \neq \beta} \sum_{i\alpha j\beta} \left(\frac{d\phi_{ij}^{LJ}}{dr_{ij}} + \frac{d\phi_{ij}^C}{dr_{ij}} \right) \frac{x_{\alpha\beta} x_{ij} + y_{\alpha\beta} y_{ij} + z_{\alpha\beta} z_{ij}}{3r_{ij}} \right\rangle \qquad (2.28)$$

The long-range contribution of LJ pair interactions is then

$$P_{LR}^{LU} = 8\pi \sum_{ij} \rho_i \rho_j \varepsilon_{ij} \left(\frac{4\sigma_{ij}^{12}}{9r_c^9} - \frac{2\sigma_{ij}^6}{3r_c^3} \right) \tag{2.29}$$

With these expressions we can obtain the grand-potential of a phase confined in a slit and the pressure of a bulk phase. Where metastable phases exist the equilibrium phase is that with the lowest grand-potential (the highest pressure or average transverse pressure if it is a bulk phase or planar phase, respectively).

The angle brackets imply an average obtained from an ensemble of microscopic states. During a Monte Carlo simulation microscopic quantities of interest are calculated and stored at regular intervals. Since it is impossible to generate all members of the ensemble, ensemble averages will always be subject to statistical uncertainties even if there are no systematic errors. The required length of a simulation will depend upon the magnitude of fluctuations in a quantity of interest and the associated level of statistical error that is deemed satisfactory. The statistical error, v, in a series, η, of uncorrelated values, A_k is [4,5]

$$v = \sqrt{\frac{1}{\eta} \sum_{k=1}^{\eta} (A_k - \langle A \rangle)^2} \tag{2.30}$$

This expression must be divided by $\delta^{1/2}$ if, on average, blocks of length δ of the series are correlated [4,5]. This means that when fluctuations in a quantity are slow the length of the simulation must be increased to achieve a satisfactory level of statistical error.

Systematic errors are often caused by inefficient sampling of microstates, or poor ergodicity. These errors occur when the sampled microstates are not statistically representative of a single thermodynamic state. For example, near the critical temperature of a bulk-fluid, the liquid and gas phases in a Gibbs ensemble simulation can "swap" boxes and so neither box can represent one phase only. Alternatively, a poor choice of simulation move might result in rejection of the overwhelming majority of moves. This can occur in both Gibbs and Grand-canonical ensemble simulations of dense phases where it becomes increasingly unlikely that inter-box, creation and annihilation moves will be accepted as the density of a phase increases. This type of ergodicity deficiency has been called quasi-ergodicity [46].

For the case when more than one thermodynamic state is sampled, a histogram of the relevant order-parameter (for example, the density) will reveal more than one statistically significant peak. When simulating bulk fluid phases, as in the Gibbs ensemble, it can be shown that in the thermodynamic limit the locations of these peaks correspond to the equilibrium gas and liquid densities. Thus a histogram analysis provides a valuable complement to

Equation 2.31 (for a much more detailed discussion, see Ref. [6], chapter 6). Various approaches have been developed to combat quasi-ergodicity including cavity biased [47] and configurational biased sampling [48].

2.6   *Phase coexistence results*

Our goal is to fine-tune molecular models of N_2, CO, CH_4 and CO_2 to achieve good agreement between coexisting liquid and gas densities and pressures obtained from Gibbs ensemble simulation and experimental data [33–40]. We perform Gibbs ensemble simulations with N, V and T held constant and for which we first need to choose appropriate values. Clearly, we must set T to be between the appropriate experimental bulk critical point temperature and triple-point temperature. The choice of N and V is less straightforward. V should be chosen so that the instantaneous box length side, L, is always greater than twice the cut-off length, r_c. N is chosen so that the volumes of the two simulation boxes are approximately equal at equilibrium. L determines the maximum spatial correlation length obtainable by a system and this in turn affects the location of the bulk critical point. So N, V and the critical-point are affected by our choice of cut-off.

We set $r_c = 1.5$ nm and employ long-range LJ corrections [49] in all simulations. We calculate the pressure of each phase according to Equations 2.29 and 2.30 and calculate statistical errors from Equation 2.30. The best-fit molecular models are described in Table. 2.1. The liquid and gas coexisting densities and pressures are presented in Figure 2.2a–d. We note that, rather than performing additional Gibbs ensemble simulations for CH_4 in this work, we have instead fine-tuned the model parameters by fitting an equation-of-state [50] (EOS) for the Lennard–Jones fluid to the reference coexistence data at several temperatures below the critical temperature. We use the EOS of Nezbeda and Kolafa [50], which is obtained by fitting to simulation results from a wide range of sources and is thought to accurately predict coexistence pressures and densities.

Table 2.1 Model parameters for gas–gas interactions

Parameter	N_2	CH_4	CO	CO_2
σ_{ff}(nm)	0.334	0.373	C:0.349	C:0.275
			O:0.313	O:0.3015
ε_{ff}/k_B(K)	34.7	147.5	C:22.8	C:28.3
			O:63.5	O:81.0
l_x (nm)	±0.05047	0	C:+0.056	C:0
			O:−0.056	O:±0.1149
l_q (nm)	±0.0847	0	C:+0.056	C:0
	±0.1044		O:−0.056	O:±0.1149
$q(e)$	0.373	0	C:0.0203	C:0.6512
	−0.373		O:−0.0203	O:−0.3256

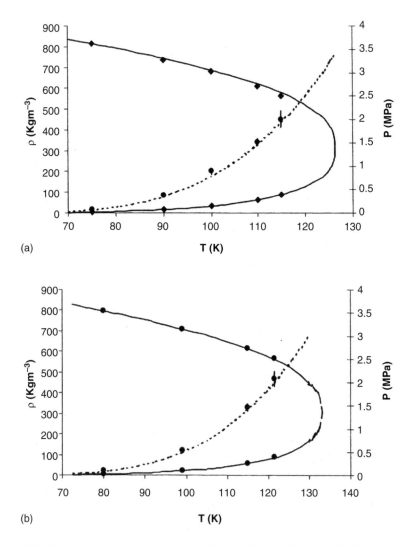

Figure 2.2 Coexistence properties of: (a) N_2, (b) CO, (c) CH_4 and (d) CO_2. Lines are experimental data (see text) and symbols are results from Gibbs ensemble simulations except for (c) where symbols are results from the EOS of Kolafa and Nezbeda [50] for a LJ fluid. All model parameters are given in Table 2.1. The experimental density data has been extrapolated to the critical point in (b) (long-dashed line).

Because of the relatively large cut-off used our simulations generally hold relatively large numbers of molecules. These molecular models might not be the most efficient computationally since it is possible that effective molecular models can be found which employ a smaller cut-off and fit the reference data equally well.

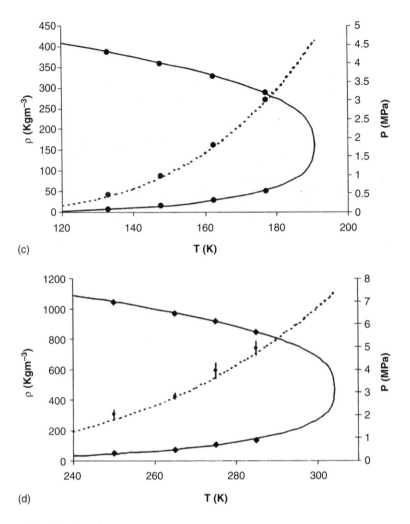

(c)

(d)

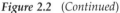

Figure 2.2 *(Continued)*

2.7 Isotherm results

The next step is to fine-tune effective molecular models of N_2, CO, CH_4 and CO_2 adsorbed on graphitic surfaces to achieve good agreement between adsorption isotherms obtained from Grand-canonical ensemble simulation and experimental data on reference materials. We perform Grand-canonical ensemble simulations with μ, A, w and T held constant. Our choice for μ, w and T is determined by available experimental data. We need to choose an appropriate value for $A = L^2$. Clearly, L should always be greater than twice the cut-off length, r_c. Also, L determines the maximum transverse spatial correlation length obtainable by a system and this in turn affects the nature of critical phenomena in adsorbed fluids, such as in wetting films.

Table 2.2 Model parameters for gas–solid surface interactions

Parameter	N_2	CH_4	CO	CO_2
σ_{sf}(nm)	0.337	0.365	C:0.3445	C:0.308
			O:0.3265	O:0.321
ε_{sf}/k_B(K)	26.0	54.3	C:21.1	C:23.8
			O:35.2	O:39.2
E_{ss}/k_B(K)	19.5	20	C:19.5	C:19
			O:19.5	O:19

In previous work [30,31] we calibrated appropriate values for ε_{ss}, for the interaction of N_2 at 77 K and N_2, CH_4 and CO_2 at 298 K with a graphitic surface. We use the same values for CO_2 on graphite at 273 K as for 298 K. We take σ_{ss} to be 0.34 nm, a commonly used value [27,51]. The individual site-surface interaction parameters, ε_{is}, and σ_{is}, are recovered with the Lorentz-Berthelot rules (3). The appropriate values for ε_{ss} and σ_{ss}, for the interaction of CO at 298 K with a graphitic surface are assumed to be identical to those for N_2 at 298 K. To be consistent with our Gibbs ensemble calculations we use the same molecular models (described in Table 2.1) and employ long-range corrections for LJ interactions only. The gas-surface interaction parameter values are presented in Table 2.2.

Figure 2.3a–d shows adsorption isotherm databases for N_2 at 77 K up to 1 bar, CO and CH_4 at 298 K and CO_2 at 273 K in graphitic slit pores. We calculate bulk pressures using Equations (2.28) and (2.29). At these temperatures N_2 is significantly sub-critical, CO and CH_4 are significantly super-critical and CO_2 is marginally sub-critical. Because of this temperature range the databases in Figure 2.3a–d show a wide range of phenomena. All the databases show adsorption generally increasing with pressure. The CO and CH_4 databases exhibit high adsorption for narrow pore widths indicative of the strongly attractive nature of the graphitic pore walls. The CH_4 database also shows a secondary maximum for adsorption at high pressure in pores able to accommodate two layers of adsorbate. This secondary maximum appears as a slight bump in the CO database, which is more supercritical than CH_4 at 298 K. But both these databases are quite featureless for higher pore widths.

The CO_2 database contains much more information than the CO and CH_4 databases. Adsorption in pores that can accept one layer of fluid only is almost "flat" at high pressure indicating that these pores are nearly saturated. The secondary maximum indicating two adsorbed layers extends to wider pores and capillary condensation is observed for the widest pores at pressures close to saturation.

Nitrogen at 77 K is closer to its triple point temperature (63 K) than its critical temperature (126 K). This is reflected in the complexity of the N_2 (77 K) database. This sensitivity of the pore density to pore width makes N_2 at 77 K an attractive choice for pore size characterisation studies for a wide range of materials. We see that the narrowest pores are saturated with N_2 even for very low under-saturated pressures. For wider slits we see capillary condensation for a wide range of slit widths. For wide slits a N_2 monolayer forms at about $P = 0.001$ bar prior to condensation at higher pressure. However, capillary condensation appears to vanish for 1.2 nm $< w <$ 1.7 nm and then re-appear for 1.0 nm $< w <$ 1.2 nm before vanishing again in smaller pores. This apparent "re-entrant" capillary condensation is probably caused by packing effects that disrupt condensation for 1.2 nm $< w <$ 1.7 nm and enhance it for 1.0 nm $< w <$ 1.2 nm. This phenomenon has been observed before [17] in DFT and simulations studies of spherical N_2 molecules. Packing effects are also responsible for the oscillations in average pore density with slit width in the condensed region of the database.

Each database result is obtained with a Grand-canonical ensemble simulation initialised with zero molecules. Due to the high free-energy barrier between gas-like (monolayer) and liquid-like (capillary condensed) states such simulations are unlikely to sample microstates corresponding to liquid-like states if the bulk pressure is too close to the capillary condensation transition pressure, P_{cc}. This means that we need to perform additional simulations initialised with liquid-like configurations to determine the properties of the liquid-like branch of the isotherm in the region. Coexisting gas-like and liquid- like states can then be determined by calculating when the grand potential (or the average transverse pressure) on each isotherm branch is equal [52]. To provide an example of this procedure we have located P_{cc} for $w = 2.512$ nm. We calculate the grand-potential by integrating the Gibbs adsorption Equation 2.22 along each branch. The constant of integration for each branch is determined at a single point using the virial Equations 2.26 and (2.27). Figure 2.4 shows the results of these calculations and also verifies that the Gibbs adsorption and virial routes to the grand potential are consistent. We find that $P_{cc} = 0.2 \pm 0.05$ bar for N_2 at 77 K in a graphitic slit of width $w = 2.512$ nm. Figure 2.5 shows the gas-like and liquid-like singlet densities (density profiles) for N_2 molecule centres at $P = 0.019$ bar. It is clear that upon condensation the density in the central region of the slit attains liquid-like values. We can also see that the N_2 layers closest to the slit walls are effectively separate from the rest of the fluid.

The N_2 database is not as accurate in the saturated region of isotherms with $w < 1.2$ nm as it is for $w \geq 1.2$ nm because our simulations exhibit quasi-ergodicity in the narrower pores. To illustrate this point we have repeated calculation of the database for $P = 0.01$ bar and a range of values for $w \leq 1.2$ nm using alternative initial configurations. These initial configurations are generated from the final configuration of simulations in which the N_2—surface interaction strength is gradually reduced from a very high value, to the calibrated value in Table 2.2. So these initial configurations are "over-dense." After simulation of a further 2 million attempted moves, we find that the average pore density is higher when using an "over-dense" initial configuration compared to an "empty" initial configuration for $w < 1.2$ nm. This indicates quasi-ergodicity for $w < 1.2$ nm resulting from the low probability of acceptance of creation and annihilation moves. We have also performed further simulations with "average-density" initial configurations. The results of these simulations for

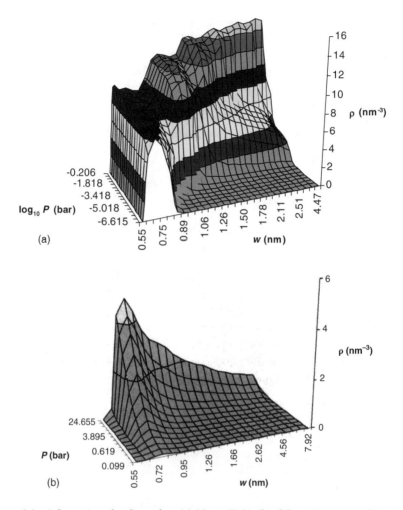

Figure 2.3 Adsorption database for: (a) N_2 at 77 K, (b) CO at 298 K, (c) CH_4 at 298 K and (d) CO_2 at 273 K, in graphitic slit pores from Grand-canonical ensemble simulation. Note that P and w are shown on logarithmic scales, except for (a) where $\log_{10} P$ (bar) is shown on a uniform scale.

$w = 0.7$, 1.0 and 1.2 nm are presented in Figure 2.6a,b. They show that equilibrium is attained for $w = 1.2$ nm but not for $w < 1.2$ nm. We estimate from these figures that the average pore density in the saturated region of the database for $w < 1.2$ is in error by about 5–10%. It is possible that methods such as the cavity biased method [47] will improve equilibration for $w < 1.2$ nm. However, the location of the condensation transition for $w < 1.2$ nm is outside of the quasi-ergodically limited region. This means that we can determine the grand-potential and the location of the phase transition for $w < 1.2$ nm. We have performed further simulations at $P = 0.019$ mbar for N_2 in graphitic slit pores with $w = 1.0$ nm. It is clear from these simulations that

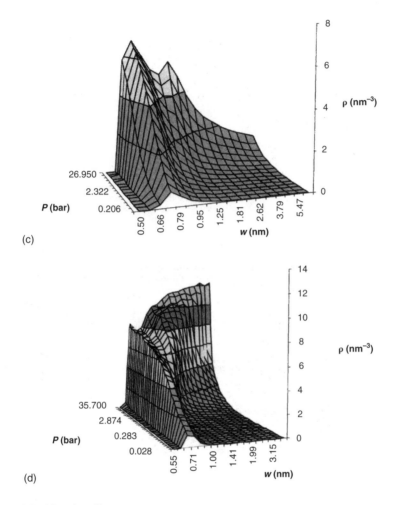

Figure 2.3 (*Continued*).

capillary condensation for $w < 1.2$ nm occurs from a gas-like state to a liquid-like state, albeit with much reduced dimensionality. We are not observing capillary freezing in these slit pores, although it is conceivable that N_2 does freeze in these pores at higher pressure. Figure 2.7a,b shows "snapshots" of gas-like and liquid-like metastable states from these simulations.

2.8   Characterisation

The characterisation of porous materials usually involves an approximate solution of the adsorption integral

$$V(P) = A \int f(w)v(w, P)dw \qquad (2.31)$$

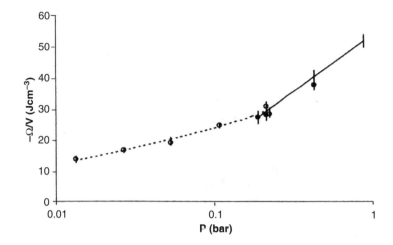

Figure 2.4 Grand potential density isotherms for N_2 at 77 K in a graphitic slit of width 2.512 nm. Lines are calculated from the Gibbs adsorption Equation 2.22, symbols from the virial Equations (2.26) and (2.27). The dashed line and open symbols indicate gas-like (monolayer) states while the solid line and filled symbols indicate liquid-like (condensed) states. Pressure is on a logarithmic scale.

where $V(P)$ is the experimentally determined excess volume of adsorbate (at STP) per gram of material, $f(w)$ is the required pore size distribution and $v(w,P)$ is the excess average density of adsorbate at pressure P in a pore of size w. The integral is overall pore sizes, w. Equation 2.31 is a Fredholm

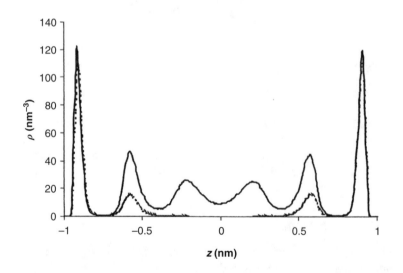

Figure 2.5 Singlet density of N_2 molecule centers at 77 K in a graphitic slit of width 2.512 nm at $P = 0.19$ bar. The solid line is a liquid-like state, the dotted line is a gas-like state and z is relative to the slit centre.

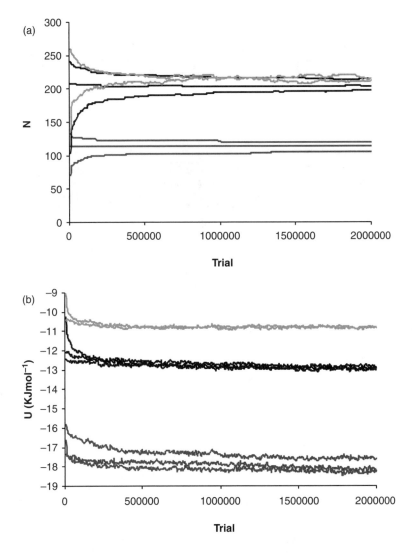

Figure 2.6 (a) Evolution of the number of molecules in Grand-canonical ensemble simulations of N_2 at $P = 0.01$ bar in graphitic slit pores with $w = 0.7$ nm (dark gray lines), 1.0 nm (black lines) and 1.2 nm (light gray lines). Lines with the same colour indicate simulations initialised with different states. (b) As for Figure 2.6a except that the evolution of the average energy per molecule is shown.

equation of the first kind, and as such it can present many difficulties. Nevertheless, several methods for solving Equation 2.31 are known including best-fit methods [53,54] and matrix methods [55,56]. The best-fit methods are essentially trial-and-error methods where very many trial functions are tested, with the best-fit trial function taken as the solution. They can employ optimisation procedures to direct the trial function selection towards better

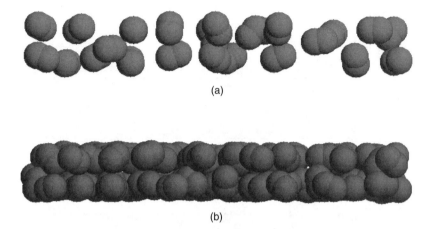

(a)

(b)

Figure 2.7 (a) A gas-like configuration of N_2 at $P = 0.019$ mbar (close to the capillary condensation pressure) in a graphitic slit with $w = 1.0$ nm. (b) As for Figure 2.7a except that a liquid-like configuration is shown.

solutions. The matrix methods amount to solving a system of linear equations by matrix inversion. With both methods, additional constraints are often required to force more physically appealing or acceptable solutions, including constraints on the smoothness of the solution function and the range of w. Any solution method for Equation 2.31 requires that the kernel, $v(w, P)$, is known. The first step is to identify a pore geometry and associated measure, w. With the standard idealised carbon slit-pore model $f(w)$ describes a poly-disperse array of slit pores. Given a fixed geometry, the function v must be calculated for all relevant values of w and P. The data presented in Figure 2.3 constitute $v(w,P)$ for each gas at particular temperatures.

From Figure 2.3 it is clear that at 293 K carbon-dioxide adsorption isotherms simulated up to pressures of 30 bar in slit pores are sensitive to slit width. It follows that for our model polydisperse slit pore material the predicted total isotherm will be sensitive to small variations in the PSD. As a consequence the PSD obtained by inverting Equation 2.31 will be constrained by the experimental isotherm. Therefore at room temperature carbon-dioxide will be a sensitive probe of the PSD of porous materials if measurements are made up to the saturation pressure. Carbon-monoxide and methane are super-critical at 298 K; the isotherms are only weak functions of pore width, and hence they are not as sensitive as carbon-dioxide as probes of the microstructure. Clearly, nitrogen isotherms at 77 K (Figure 2.3a) are the most sensitive to changes in pore width. However a significant body of theoretical and experimental evidence [57] points to the fact that experimental studies are hampered by very slow diffusion of N_2 into these materials. As discussed above, the database for nitrogen in the important range $w < 1.2$ nm is likely to be inaccurate due to quasi-ergodicity. In this case, with systematic errors in both experimental and modelling data, the pore size distribution obtained from a nitrogen isotherm at 77 K using Equation 2.31 is likely to be unreliable for microporous carbon materials.

2.9 Summary

We have given an overview of current modelling procedures used to predict the adsorption of a range of gases in graphitic pores complementary to that of Nicholson [58]. The adsorbed fluids display a wide variety of adsorption behaviours in small pores depending on their interaction potentials and the temperature. From our data we see that at or near room temperature the CO_2 database of isotherms contains much more information than the CO and CH_4 databases. Therefore at room temperature carbon-dioxide will be a sensitive probe of the pore size distribution of porous materials if measurements are made up to the saturation pressure. Nitrogen at 77 K is closer to its triple point temperature (63 K) than its critical temperature (126 K). This is reflected in the complexity of the N_2 (77 K) database. This sensitivity of the pore density to pore width in principle makes nitrogen at 77 K an attractive choice for pore size characterisation studies for a wide range of materials. However both the experimental isotherms and the simulation database are likely to be inaccurate due to the possibility that the equilibrium state of nitrogen in the smallest pores or near pore junctions is solid. Clearly the safest choice is to characterise nanoporous materials using carbon dioxide isotherms at room temperature. Given accurate potentials the techniques discussed here can be used to predict adsorption selectivity both for single pores [59–61] and for an assembly of pores representing the pore size distribution of a real material [62]. An interesting extension of the present work will be to consider the phase behaviour, structure and transport properties of gas mixtures containing water in graphitic nanopores building on the work of Nicholson and colleagues [63].

Acknowledgments

It is a great pleasure for us to acknowledge many years of useful discussions with David Nicholson. As is clear from the many references to his work in the present chapter he has been a pioneer in the application of molecular simulation methods to the study of physical adsorption. We thank EPSRC for support through grant GR/M94427.

References

1. Gregg, S.J. and Sing, K.S.W. (1991) *Adsorption, Surface Area and Porosity*, 2nd ed. (Academic Press, New York).
2. Rouquerol, F., Rouquerol, J. and Sing, K. (1999) *Adsorption by Powders and Porous Solids* (Academic Press, New York).

3. See Evans, R. in (1992) *Fundamentals of Inhomogeneous Fluids,* Henderson, D. Ed, (Dekker, New York), for a detailed survey.

4. Allen, M.P. and Tildesley, D.J. (1987) *Computer Simulation of Liquids* (Clarendon Press, Oxford).

5. Frenkel, D. and Smit, B. (1996) *Understanding Molecular Simulation: From Algorithms to Applications* (Academic Press, New York).

6. Allen, M.P. and Tildesley, D.J., Eds, *Computer Simulation in Chemical Physics NATO ASI Series C: Mathematical and Physical Sciences,* (Kluwer Academic Publishers, Dordrecht) Vol. 397, 1993.

7. Rowley, L.A., Nicholson, D. and Parsonage, N.G. (1976) "Grand ensemble Monte Carlo studies of physical adsorption I. Results for multilayer adsorption of 12-6 argon in the field of a plane homogeneous solid," *Mol. Phys.* 31, 365.

8. Metropolis, N., Rosenbluth, A.W., Rosenbluth, M.N., Teller, A.H. and Teller, E., "Equation of state calculations by fast computing machines," (1953) *J. Chem. Phys.* 21, 1087.

9. Norman, G.E. and Filinov, V.S. (1969) "Investigations of phase transitions by a Monte Carlo method," *High Temp. (USSR)* 7, 216.

10. Rowley, L.A., Nicholson, D. and Parsonage, N.G. (1975) "Monte Carlo grand canonical ensemble calculations in a gas–liquid transition region for 12–6 argon," *J. Comput. Phys.* 17, 401.

11. Adams, D.J. (1975) "Grand-canonical ensemble Monte Carlo for a Lennard–Jones fluid," *Mol. Phys.* 29, 307.

12. Israelachvili, J. and Gourdon, D. (2001) "Liquids—putting liquids under molecular-scale confinement," *Science* 292, 867.

13. Panagiotopoulos, A.Z. (1987) "Direct determination of phase coexistence properties of fluids by Monte Carlo simulation in a new ensemble," *Mol. Phys.* 61, 813.

14. Panagiotopoulos, A.Z., Quirke, N., Stapleton, M. and Tildesley, D.J. (1988) "Phase equilibria by simulation in the Gibbs ensemble: alternative derivation, generalization and application to mixture and membrane equilibria," *Mol. Phys.* 63, 527.

15. Smit, B., de Smedt, P.H. and Frenkel, D. (1989) "Computer simulations in the Gibbs ensemble," *Mol. Phys.* 68, 931.

16. Panagiotopoulos, A.Z. (1987) "Adsorption and capillary condensation of fluids in cylindrical pores by Monte Carlo simulation in the Gibbs ensemble," *Mol. Phys.* 62, 701.

17. Lastoskie, C., Gubbins, K.E. and Quirke, N. (1993) "Pore size heterogeneity and the carbon slit pore: a density functional theory model," *Langmuir* 9, 2693.

18. Cracknell, R.F., Nicholson, D., Tennison, S.R. and Bromhead, J. (1996) "Adsorption and selectivity of carbon-dioxide with methane and nitrogen in slit-shaped carbonaceous micropores: simulation and experiment," *Adsorption* 2, 193.

19. Nicholson, D. and Gubbins, K.E. (1996) "Separation of carbon-dioxide-methane mixtures by adsorption: effects of geometry and energetics on selectivity," *J. Chem. Phys.* 104, 8126.

20. Kaneko, K., Cracknell, R.F. and Nicholson, D. (1994) "Nitrogen adsorption in slit pores at ambient-temperatures—comparison of simulation and experiment," *Langmuir* 10, 4606.

21. Whitehouse, J.S., Nicholson, D. and Parsonage, N.G. (1983) "A grand ensemble Monte Carlo study of krypton adsorbed on graphite," *Mol. Phys.* 49, 829.

22. Nicholson, D. and Parsonage, N.G. (1986) "Simulation studies of the fluid-monolayer transition in argon adsorbed on graphite," *J. Chem. Soc., Faraday Trans.* 2, 1657.

23. Ewald, P. (1921) "Die Berechnung optischer und elektrostatischer Gitter-potemtiale," *Ann. Phys.* 64, 253.

24. Heyes, D.M. (1981) "Electrostatic potentials and fields in infinite point charge lattices," *J. Chem. Phys.* 74, 1924.

25. Heyes, D.M. (1983) "MD incorporating Ewald summations on partial charge polyatomic systems," *CCP5 Quarterly* 8, 29.

26. Barker, J.A. and Watts, R.O. (1973) "Monte Carlo studies of the dielectric properties of water-like molecules," *Mol. Phys.* 26, 789.

27. Steele, W.A. (1973) "The physical interactions of gases with crystalline solids," *Surf. Sci.* 36, 317.

28. Hanson, J.P. and McDonald, I.R. (1986) *Theory of Simple Liquids,* 2nd ed. (Academic Press, New York).

29. Leckner, J. (1991) "Summation of coulomb fields in computer-simulated dis-ordered systems," *Physics A* 176, 485.

30. Sweatman, M.B. and Quirke, N. (2001) "Characterization of porous materials by gas adsorption at ambient temperatures and high pressure," *J. Phys. Chem. B* 105, 1403.

31. Sweatman, M.B. and Quirke, N. (2001) "The characterisation of porous ma-terials by gas adsorption: comparison of nitrogen at 77 K and carbon-dioxide at 298 K for activated carbon," *Langmuir.* 17, 5011.

32. Sokhan, V.P., Nicholson, D. and Quirke, N. (2000) "Phonon spectra in model carbon nanotubes," *J. Chem. Phys.* 113, 2007.

33. Duschek, W., Kleinrahm, R. and Wagner, W. (1990) "Measurement and corre-lation of the (pressure, density, temperature) relation of carbon-dioxide. 1. The homogeneous gas and liquid regions in the temperature range from 217 K to 340 K at pressures up to 9 MPa," *J. Chem. Therm.* 22, 827.

34. Duschek, W., Kleinrahm, R. and Wagner, W. (1990) "Measurement and cor-relation of the (pressure, density, temperature) relation of carbon-dioxide. 2. Saturated-liquid and saturated-vapour densities and the vapour pressure along the entire coexistence curve," *J. Chem. Therm.* 22, 841.

35. Pierperbeck, N., Kleinrahm, R. and Wagner, W. (1991) "Results of (pressure, density, temperature) measurements on methane and on nitrogen in the tem-perature range from 273.15 K to 323.15 K at pressures up to 12MPa using a new apparatus for accurate gas-density measurements," *J. Chem. Therm.* 23, 175.

36. Handel, G., Kleinrahm, R. and Wagner, W. (1992) "Measurements of the (pressure, density, temperature) relation of methane in the homogeneous gas and liquid regions in the temperature range from 100 K to 206 K and at pressures up to 8 MPa," *J. Chem. Therm.* 24, 685.

37. Gilgen, R., Kleinrahm, R. and Wagner, W. (1992) "Supplementary measure-ments of the (pressure, density, temperature) relation of carbon-dioxide in the homogenous region at temperatures from 220 K to 360 K and pressures up to 13 MPa," *J. Chem. Therm.* 24, 1243.

38. Nowak, P., Kleinrahm, R. and Wagner, W. (1997) "Measurement and correla-tion of the (P, rho, T) relation of nitrogen. 2. Saturated-liquid and saturated-vapour densities and vapour pressures along the entire coexistence curve," *J. Chem. Therm.* 29, 1157.

39. Wagner, W. and de Reuck, K.M. (1996) International Thermodynamic Tables of the Fluid State. 13, Methane IUPAC Chemical Data Series No. 41, (IUPAC, Oxford).

40. Din, F. (1956) *Thermodynamic Functions of Gases Ammonia, Carbon-Dioxide and Carbon-Monoxide,* (Butterworths, London) Vol. 1.

41. Callen, H.B. (1985) *Thermodynamics and an Introduction to Thermostatistics,* 2nd ed. (Wiley, New York).

42. Rowlinson, J.S. and Widom, B. (1989) *Molecular Theory of Capillarity* (Clarendon Press, Oxford).

43. Nicholson, D. and Parsonage, N.G. (1982) *Computer Simulation and the Statistical Mechanics of Adsorption* (Academic Press, New York).

44. Henderson, J.R. in (1992) *"Fundamentals of Inhomogeneous Fluids,"* Henderson, D., Ed, (Dekker, New York).

45. Alejandre, J., Tildesley, D.J. and Chapela, G.A. (1995) "Molecular dynamics simulation of the orthobaric densities and surface tension of water," *J. Chem. Phys.* 102, 4574.

46. Valleau, J.P. and Whittington, S.G. (1986) *Statistical Mechanics* (Plenum Press, New York).

47. Mezei, M. (1980) "A cavity biased $(TV\mu)$ Monte Carlo method for the computer simulation of fluids," *Mol. Phys.* 40, 901.

48. Cracknell, R.F., Nicholson, D., Parsonage, N.G. and Evans, H. (1990) "Rotational insertion bias—a novel method for simulating dense phases of structured particles, with particular application to water," *Mol. Phys.* 71, 931.

49. Rowley, L.A., Nicholson, D. and Parsonage, N.G. (1978) "Long-range corrections to Grand-canonical ensemble Monte Carlo calculations for adsorption systems," *J. Comput. Phys.* 26, 66.

50. Kolafa, J. and Nezbeda, I. (1994) "The Lennard–Jones fluid—an accurate analytic and theoretically based equation of state," *Fluid Phase Equilibr.* 100, 1.

51. Hirschfelder, J.O., Curtiss, C.F. and Bird, R.B. (1964) *Molecular Theory of Gases and Liquids* (Wiley, New York).

52. Walton, J.P.R.B. and Quirke, N. (1989) "Capillary condensation: A molecular simulation study," *Mol. Sim.* 2, 361.

53. Seaton, N.A., Walton, J.P.R.B. and Quirke, N. (1989) "A new analysis method for the determination of the pore size distribution of porous carbons from nitrogen adsorption measurements," *Carbon* 27, 853.

54. Scaife, S., Kluson, P. and Quirke, N. (1999) "Characterisation of porous materials by gas adsorption," *J. Phys. Chem. B* 104, 313.

55. Ravikovitch, P.I., Vishnyakov, A., Russo, R. and Neimark, A.V. (2000), "Unified approach to porous size characterization of microporous carbonaceous materials from N-2, Ar, and CO_2 adsorption isothermis" *Langmuir* 16, 2311.

56. Gusev, V.Y., O'Brien, J.A. and Seaton, N.A. (1997), "A self-consistent method for characterization of activated carbons using super critical adsorption and grand canonical Monte Carlo simulations," *Langmuir* 13, 2815.

57. Garcia-Martinez, J., Cazorla-Amoros, D. and Linares-Solano, A. (2000), "Further evidences of the usefulness of CO_2 adsorption to characterise microporous solids," *Stud. Surf. Sci. Catal.* 128, 485.

58. Nicholson, D. (1996) "Using computer simulation to study the properties of molecules in micropores," *J. Chem. Soc., Faraday Trans.* 92, 1.

59. Cracknell, R.F. and Nicholson, D. (1995) "Adsorption of gas mixtures on solid surfaces, theory and computer simulation," *Adsorption* 1, 16.

60. Cracknell, R.F., Nicholson, D. and Quirke, N. (1994) "A Grand-canonical Monte Carlo study of Lennard–Jones mixtures in slit pores 2. Mixtures of 2 centre ethane with methane," *Mol. Sim.* 13, 161.

61. Cracknell, R.F., Nicholson, D. and Quirke, N. (1993) "A Grand-canonical Monte Carlo study of Lennard–Jones mixtures in slit pores," *Mol. Phys.* 80, 885.

62. Kluson, P., Scaife, S. and Quirke, N. (2000) "The Design of Microporous Graphitic Adsorbents for Selective Separation of Gases," *Separat. Purificat. Rev.* 20, 15.

63. Christou, N.I., Whitehouse, J.S., Nicholson, D. and Parsonage, N.G. (1981) "A Monte Carlo study of fluid water in contact with structureless walls," *R. Soc. Chem. Faraday Symp.* 16, 139.

chapter three

Effect of confinement on melting in slit-shaped pores: experimental and simulation study of aniline in activated carbon fibers

M. Śliwinska-Bartkowiak
Adam Mickiewicz University

R. Radhakrishnan
Massachusetts Institute of Technology

K.E. Gubbins*
North Carolina State University

Contents

* Corresponding author.
Reprint from *Molecular Simulation*, 27: 5–6, 2001. http://www.tandf.co.uk

3.1   Introduction

Recent molecular simulation studies for pores of simple geometry have shown a rich phase behavior associated with melting and freezing in confined systems [1–9]. A review of the simulation and experimental work in this area up to 1999 has been given by Gelb et al. [8]. The freezing temperature may be lowered or raised relative to the bulk freezing temperature, depending on the nature of the adsorbate and the porous material. In addition, new surface-driven phases may intervene between the liquid and solid phases in the pore. "Contact layer" phases of various kinds often occur, in which the layer of adsorbed molecules adjacent to the pore wall has a different structure from that of the adsorbate molecules in the interior of the pore. For materials having walls that are weakly attractive (e.g., glasses, silicas) this contact layer is usually fluid-like while the interior molecules have adopted a crystalline structure. For materials such as carbon, which has walls that are strongly attractive, the contact layer is usually crystalline while the interior layers remain fluid. These contact layer phases have been predicted theoretically, and confirmed experimentally for several systems [3,7]. In addition, for some systems in which strong layering of the adsorbate occurs (e.g., slit pore models of activated carbon fibers), hexatic phases can occur; such phases have quasi-long-ranged orientational order, but positional disorder, and for quasi-two-dimensional systems occur over a temperature range between those for the crystal and liquid phases. These are clearly seen in molecular simulations [2,8], and preliminary experiments seem to confirm these phases [7,9]. Recently it has been shown [7,10] that this apparently complex phase behavior results from a competition between the fluid–wall and fluid–fluid intermolecular interactions. For a given pore geometry and width, the phase diagrams for a wide range of adsorbates and porous solids can be classified in terms of a parameter α that is the ratio of the fluid–wall to fluid–fluid attractive interaction [7,8,10].

In addition to the strong fundamental scientific interest, an understanding of freezing in confined systems is of practical importance in lubrication, adhesion, nano-tribology and fabrication of nano-materials. The use of nano-porous materials as templates for forming nano-materials such as nano-wires and nano-tube arrays is receiving wide attention. Recent examples have included the use of track-etched pores in anodized alumina to form nano-wire/nano-tube arrays [11], carbon nano-tubes for growing nano-wires [12], and opals to obtain aligned nano-particles [13,14]; formation of the nano-material in the porous template is usually achieved by infiltration of molten material [14], vapor phase deposition [15], or electrochemical deposition [16].

In this paper we report an experimental study of the freezing behavior of aniline confined within activated carbon fibers having a pore width of approximately 1.8 nm. Dielectric relaxation spectroscopy is used to locate phase transitions, and to determine the dielectric relaxation time for confined aniline as a function of the temperature. Such relaxation times are a sensitive measure of the type of phase that is present, and the results indicate that two transitions occur on heating. We also report a molecular simulation study for a simple model of this system, in which the pores are represented as slit-shaped. Free energy calculations based on Landau theory, together with calculations of pair correlation functions, enable us to locate phase transitions and to determine the nature of the phases involved. The simulations indicate two transitions on heating, the first from a confined crystal to a hexatic phase, and the second from hexatic to liquid phase.

3.2 Experimental method

Freezing of a dipolar liquid is accompanied by a rapid decrease in its electric permittivity [17–19]. After the phase transition to the solid-state dipole rotation ceases, and the electric permittivity is almost equal to n^2, where n is the refractive index of the solid, since it arises from deformation polarisation only. Therefore, dielectric spectroscopy is suited to the investigation of melting and freezing of dipolar liquids, because significant changes occur in the system's capacity at the phase transition. Investigation of the dynamics of a confined liquid is also possible from the frequency dependence of dielectric properties of such systems. Analysis of the frequency dependence of dielectric data allows a determination of the phase transition temperature of the adsorbed substance and also of characteristic relaxation frequencies related to molecular motion in particular phases [17–19].

The dielectric relaxation method was applied to study the process of freezing and melting of a sample of aniline confined in activated carbon fibres (ACF) of type P20, obtained from the laboratory of K. Kaneko. The ACF has pores that are approximately slit-shaped, with a mean pore size $H = 1.8$ nm [6]. The complex electric permittivity, $\kappa = \kappa' + i\kappa''$, where $\kappa' = C/C_0$ is the real, and $\kappa'' = \tan(\delta)/\kappa'$ is the complex part of the permittivity, was measured in the frequency interval 300 Hz–1 MHz at different temperatures by a Solartron 1200 impedance gain analyser, using a parallel plate capacitor made of stainless steel. In order to reduce the high conductivity of the sample, which was placed between the capacitor plates as a suspension of aniline filled ACF particles in pure liquid aniline, the electrodes were covered with a thin layer of teflon. From the directly measured capacitance, C, and the tangent loss $\tan(\delta)$, the values of κ' and κ'' were calculated for the known sample geometry [3,20]. The temperature was controlled to an accuracy of 0.1 K using a platinum resistor Pt(100) as a sensor and a K30 Modinegen external cryostat coupled with a N-180 ultra-cryostat.

The aniline sample was twice distilled under reduced pressure and dried over Al_2O_3. The conductivity of purified aniline was on the order of 10^{-9} Ω^{-1} m^{-1}. The ACF material to be used in the experiment was heated to about 600 K, and kept under vacuum ($\sim 10^{-3}$ Torr) for 6 days prior to the introduction of the fluid.

For an isolated dipole rotating under an oscillating field in a viscous medium, the Debye dispersion relation is derived in terms of classical mechanics as:

$$\kappa = \kappa'_\infty + \frac{\kappa'_s + \kappa'_\infty}{1 + (i\omega\tau)}, \tag{3.1}$$

where ω is the frequency of the applied potential and τ is the orientational relaxation time of a dipolar molecule. The subscript s refers to static permittivity (low frequency limit, when the dipoles have enough time to be in phase with the applied field). The subscript ∞ refers to the high frequency limit, and is a measure of the induced component of the permittivity. The dielectric relaxation time was calculated by fitting the dispersion spectrum of the complex permittivity near resonance to the Debye model of orientational relaxation.

3.3   Molecular simulation method

We have carried out Grand Canonical Monte Carlo (GCMC) simulations for a simple model of the aniline/ACF system, consisting of a Lennard–Jones fluid adsorbed in regular slit shaped pores of pore width H. Here H is the distance separating the planes through the centers of the surface-layer atoms on opposing pore walls. The fluid–fluid interaction between the adsorbed fluid molecules is modeled using the Lennard–Jones [6,12] potential. The fluid–wall interaction is modeled using a "10-4-3" Steele potential [21].

$$\phi_{fw>}(z) = 2\pi\rho_w\varepsilon_{fw}\sigma_{fw}^2\Delta\left[\frac{2}{5}\left(\frac{\sigma_{fw}}{z}\right)^{10} - \left(\frac{\sigma_{fw}}{z}\right)^4 - \left(\frac{\sigma_{fw}^4}{3\Delta(z+0.61\Delta)^3}\right)\right] \tag{3.2}$$

Here, the σs and εs are the size and energy parameters in the LJ potential, the subscripts f and w denote fluid and wall respectively, ρ_w is the density of wall atoms, Δ is the spacing between successive layers of wall atoms, and z is the distance between the adsorbate atom and the nearest point in the wall. The carbon-carbon potential parameters and structural data were taken from Steele [21]. The values are: $\rho_w = 114$ nm^{-3}, $\sigma_{ww} = 0.34$ nm, $\varepsilon_{ww}/k = 28$ K, $\Delta = 0.335$ nm. For a given pore width H, the total potential energy from both walls is given by

$$\phi_{pore}(z) = \phi_{fw}(z) + \phi_{fw}(H-z) \tag{3.3}$$

The intermolecular potential parameters were obtained as follows [7]. The intermolecular potential parameters for aniline–aniline interactions were obtained by fitting molecular simulation data for the bulk phase melting point at 1 atm pressure to experimental data [22,23]; this gave $\varepsilon_{ff}/k = 395$ K, $\sigma_{ff} = 0.514$ nm. This effective Lennard–Jones potential accounts for the dipolar forces in a crude way through the enlarged value of the well depth parameter. Lennard–Jones interactions were also used for the fluid–wall interactions.

The parameters were estimated using the Lorentz–Berthelot combining rules; the above values were used for the carbon–carbon parameters, but for the aniline–aniline parameters we used those determined for the Stockmayer potential as fitted to second virial coefficients [24] so that the well depth parameter reflects the dispersion interaction without dipolar effects. These latter parameters for aniline were $\varepsilon_{ff}/k = 358$ K, and $\sigma_{ff} = 0.514$ nm.

The relative strength of the fluid-wall to the fluid-fluid interaction is determined by the parameter $\alpha = (2/3)\,\rho_w \varepsilon_{fw}\sigma_{fw}^2 \Delta/\varepsilon_{ff}$. In the case of aniline in activated carbon fibers, $\alpha = 1.2$. Our objective is to calculate the freezing temperatures in the confined phase to compare with the experimental results.

The simulation runs were performed in the grand canonical ensemble, fixing the chemical potential μ, the volume V of the pore, and the temperature T. A pore width of $H = 3\sigma_{ff}$ was chosen to enable comparison with our experimental results. A rectilinear simulation cell of dimensions $L \times L$ (where L equals $60\sigma_{ff}$) in the plane parallel to the pore walls was used. Typically, the system contained approximately 12,000 adsorbate molecules. The adsorbed molecules formed distinct layers parallel to the plane of the pore walls. The simulation was set up such that insertion, deletion and displacement moves were attempted with equal probability, and the displacement step was adjusted to have a 50% probability of acceptance. Thermodynamic properties were averaged over 2000 million individual Monte Carlo steps. The length of the simulation was adjusted such that a minimum of 50 times the average number of particles in the system would be inserted and deleted during a single simulation run.

The method for obtaining the free energy relies on the calculation of the Landau free energy as a function of an effective bond orientational order parameter Φ, using GCMC simulations [2]. The Landau free energy, Λ, is defined by,

$$\Lambda[\Phi] = -k_B T \ln(P[\Phi]) + \text{constant} \tag{3.4}$$

where $P[\Phi]$ is the probability of observing the system having an order parameter value between Φ and $\Phi + \delta\Phi$. The probability distribution function $P[\Phi]$ is calculated in a GCMC simulation as a histogram, with the help of umbrella sampling. The grand free energy Ω is then related to the Landau free energy by

$$\exp(-\beta\Omega) = \int d\Phi \exp(-\beta\Lambda[\Phi]) \tag{3.5}$$

The grand free energy at a particular temperature can be calculated by numerically integrating Equation 3.5 over the order parameter space. We use a two-dimensional order parameter to characterize the order in each of the molecular layers.

$$\Phi_{6j} = \left| \frac{1}{N_b} \sum_{k=1}^{N_b} \exp(i6\theta_k) \right| = |\langle \exp(i6\theta_k) \rangle j| \tag{3.6}$$

Φ_{6j} measures the hexagonal bond order within each layer j. Each nearest neighbor bond has a particular orientation in the plane of the given layer, and is described by the polar coordinate θ. The index k runs over the total number of nearest neighbor bonds N_b in layer j. The overall order parameter Φ_6 is an average of the hexagonal order in all the layers. We expect $\theta_{6j} = 0$ when layer j has the structure of a two-dimensional liquid, $\theta_{6j} = 1$ in the two dimensional hexagonal crystal phase, and $0 < \Phi_{6j} < 1$ in an orientationally ordered layer.

3.4   Results

3.4.1   Experiment

The capacitance C and loss tangent tan (δ) were measured as a function of frequency and temperature for bulk aniline and for aniline adsorbed in ACF, from which the dielectric permittivity $\kappa'(T,\omega)$ and the loss tangent $\kappa''(T,w)$ were calculated. Results of the measurements of C for bulk aniline as a function of T and at the frequency of 0.6 MHz are shown in Figure 3.1. There is a sharp increase in C at $T = 267$ K, the melting point of the pure substance, due to the contribution to the orientational polarisation in the liquid state from the permanent dipoles [17,18]. In the frequency interval studied we could only detect the low-frequency relaxation of aniline. Analysis of the Cole–Cole representation of the complex permittivity for solid aniline has shown that the relaxation observed should be approximated by a symmetric distribution of relaxation times described formally by the Cole–Cole Equation 3.1. Examples of the experimental results and the fitted curves are given in Figure 3.2(a) for the bulk solid phase at 260 K. From the plot of κ' and κ'' vs. $\log_{10}(\omega)$ the

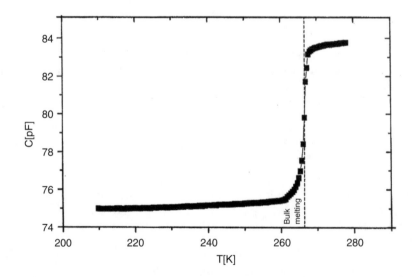

Figure 3.1 Capacitance, C, vs. temperature, T, for bulk aniline at $\omega = 0.6$ MHz.

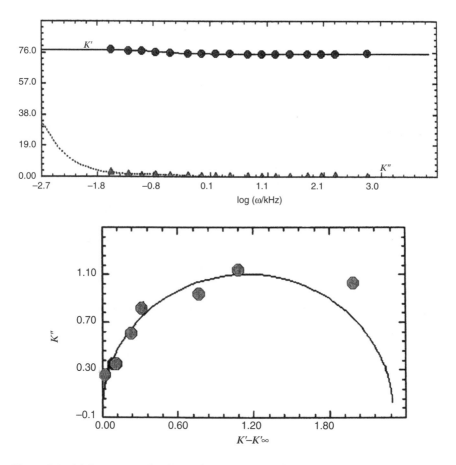

Figure 3.2   (a) Spectrum plot for aniline at 260 K. The solid and the dashed curves are fits to the real and imaginary parts of κ. (b) Representation of the spectrum plot in the form of a Cole–Cole diagram for bulk aniline at 260 K.

relaxation time can be calculated as the inverse of the frequency corresponding to a saddle point of the κ' plot or a maximum of the κ'' plot. An alternative graphical representation of the Debye dispersion equation is the Cole–Cole diagram in the complex κ plane, shown in Figure 3.2(b). Each relaxation mechanism is reflected as a semicircle in the Cole–Cole diagram. From the plot of κ'' vs. κ', the value of τ is given as the inverse of the frequency at which κ'' goes through a maximum.

In Figure 3.3 the variation of the relaxation time with temperature is presented for bulk aniline, as obtained from fitting Equation 3.1 to the dispersion spectrum. In the solid phase (below 267 K), our measurements showed a single relaxation time of the order of 10^{-3}–10^{-4} s in the temperature range from 240 to 267 K. The liquid branch above 267 K has rotational relaxation times of the order of 10^{-11} s [17,18]. This branch lies beyond the possibilities of our analyser. In the presence of dipolar constituents, one or more absorption regions are

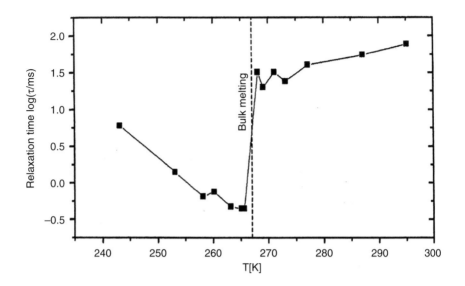

Figure 3.3   Dielectric relaxation time, τ, vs. temperature for bulk aniline.

present, not all of them necessarily associated with the dipolar dispersion. At the lowest frequencies (especially about 1 KHz), a significantly large κ'' value arises from the conductivity of the medium and interfacial (Maxwell–Wagner) polarisation is found if the system is not in a single homogeneous phase. For aniline, a homogeneous medium whose conductivity is of the order of 10^{-9} $\Omega^{-1}\,m^{-1}$, the absorption region observed for the frequencies 1–10 KHz is related to the conductivity of the medium. The Joule heat arising from the conductivity contributes to a loss factor κ'' (conductance) so the value at low frequency is: $\kappa''(\text{total}) = \kappa''(\text{dielectric}) + \kappa''(\text{conductance})$, and the system reveals the energy loss in processes other than dielectric relaxation. In Figure 3.3 the branch above 267 K, corresponding to relaxation times of the order of 10^{-2} s, characterises the process of adsorption related to the conductivity of the medium. This branch is characteristic of the liquid phase and is a good indicator of the appearance of this phase.

The behavior of C vs. T for aniline in ACF at a frequency of 0.6 MHz is shown in Figure 3.4. The sample was introduced between the capacitor plates as a suspension of ACF in pure aniline. The sharp increase in C at 267 K seen in Figure 3.4(a) is due to the bulk solid–liquid transition, and we do not observe additional changes of C characteristic of phase transitions for temperatures lower than the bulk melting point. The behavior of C vs. T for aniline in ACF at temperatures in the range 290–340 K for frequencies of 0.01, 0.1 and 1 MHz is shown in Figure 3.4(b). In this temperature range we observe two sudden changes in C that are not observed in the case of bulk aniline, and must be related to changes in the aniline confined within ACF. These changes occur at 298 and 324 K, and indicate

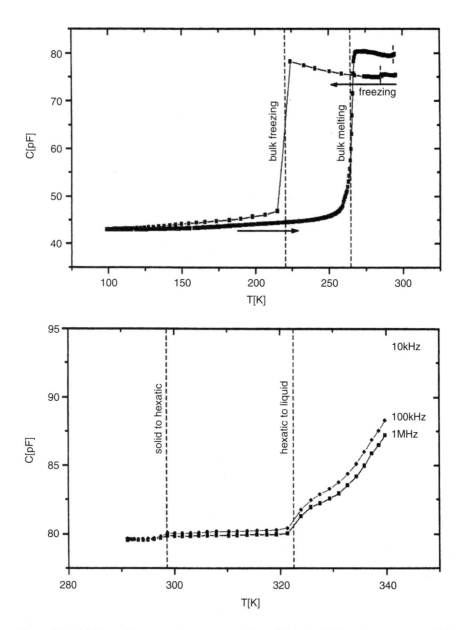

Figure 3.4   (a) Capacitance vs. temperature for aniline in ACF at frequency $\omega = 0.6$ MHz at lower temperatures, (b) C vs. T for aniline in ACF at different frequencies for temperatures 290–340 K.

phase or structural transitions in the confined phase. The latter change is very similar to that corresponding to the melting of a dipolar liquid placed in porous glass [3,20], where a significant increase in C indicated a phase transition to the liquid phase. These results suggest that the melting process

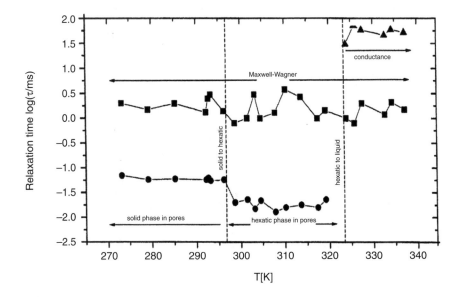

Figure 3.5   Dielectric relaxation times, τ vs. T for aniline in ACF.

of aniline confined in ACF has two steps: from solid to intermediate phase (at 298 K), and from intermediate phase to liquid at 324 K.

The characteristic relaxation times related to molecular motion in particular phases are presented at Figure 3.5, where the behavior of the relaxation times as a function of temperature for aniline in ACF is depicted. For the temperature range 273–340 K there are several different kinds of relaxation. The larger component of the relaxation time of the order 10^{-2} s is related to the conductivity of aniline in pores and testifies to the presence of a liquid phase in the system. This component appears at 324 K, where the second transition was observed in Figure 3.4(b). A relaxation time related to the Maxwell–Wagner polarisation, of the order 10^{-3} s and characteristic of interfacial polarisation, is observed over the whole temperature range. At temperatures below 298 K we observe a relaxation time of the order 10^{-4} s, which is typical of the aniline solid (crystal) phase.

Above this temperature, in the range 298–324 K, a branch of relaxation time of the order 10^{-5} s appears. This branch can be related to a Debye-type dispersion in the intermediate phase, which could be a hexatic phase [20].

3.5   Simulation results

The reduced melting temperature of the bulk Lennard–Jones fluid at 1 atm. pressure is $k_B T_f / \varepsilon_{ff} = 0.682$. For our model of aniline this corresponds to a melting temperature of 269 K, very close to the experimental value of 267 K. During the course of our GCMC simulations the distribution function $P[\Phi]$

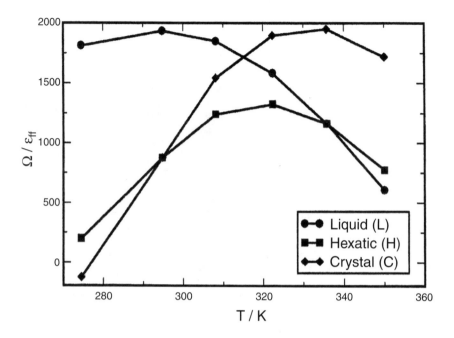

Figure 3.6   Grand free energy vs. temperature for the three phases observed, from molecular simulation.

was calculated as a function of Φ, and hence the Landau free energy and grand free energy were obtained from Equations 3.4 and 3.5, respectively. At most temperatures, the Landau free energy plots versus Φ showed three local minima, corresponding to three phases. At a given temperature, one of these minima was the global one, indicating the thermodynamically stable phase at that temperature; at temperatures at which two of the phases were in thermodynamic equilibrium two of these minima had the same Landau free energy. The state conditions of phase coexistence were determined by requiring the grand free energies of the two confined phases to be equal.

The resulting grand free energy curves for the three phases are shown in Figure 3.6. Two thermodynamic phase transitions are observed, one at 296 K and the other at 336 K. The phase transitions are seen to be first order, at least for this size of simulation box. These free energies give no information about the nature of the phases involved. However, we note that in these confined systems, because of the slit-shaped pore and narrow pore width, the adsorbate molecules are confined to layers that are quasi-two-dimensional systems. Nelson and Halperin [25] proposed the KTHNY (Kosterlitz–Thouless– Halperin–Nelson–Young) mechanism for the melting of a two dimensional crystal in two dimensions, which involves two transitions of the Kosterlitz– Thouless [26] type. The first is a transition from the two-dimensional crystal phase, with quasi-long range positional order and long-range orientational order, to a hexatic phase with long-range orientational order but positional disorder;

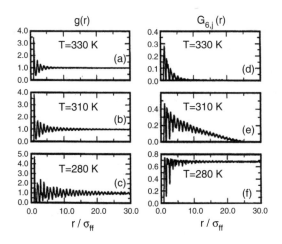

Figure 3.7 Pair correlation functions in the two confined molecular layers of aniline in ACF from simulation: (a) g(r) in the liquid phase; (b) g(r) in the hexatic phase; (c) g(r) in the crystal phase; (d) $G_{6j}(r)$ in the liquid phase; (e) $G_{6j}(r)$ in the hexatic phase; (f) $G_{6j}(r)$ in the crystal phase.

the second transition is from the hexatic phase to a liquid (having neither long range positional or orientational order). The hexatic phase was first observed experimentally in electron diffraction experiments on liquid crystal-line thin films [27]. Molecular simulation studies by Zangi and Rice [28] for quasi-two dimensional films in which some out of plane motion is permitted showed two phase transitions as proposed by the KTHNY mechanism, with the hexatic phase as the intermediate one between crystal and liquid.

In order to determine the nature of the phases shown in Figure 3.6 we calculated the in-plane pair positional and orientational correlation functions at each temperature, since these provide a clear signature of fluid, crystal and hexatic phases. Typical results are shown in Figure 3.7 for temperatures in the stable regions of the three different phases. In this figure g(r) is the usual two-dimensional in plane pair correlation function, or radial distribu-tion function. The orientational pair correlation function, $G_{6j}(r)$, for the con-fined molecular layer *j* is defined as

$$G_{6j}(r) = \langle \Phi_{6j}^{*}(0)\Phi_{6j}(r)\rangle \qquad (3.7)$$

At the highest temperature, 330 K, the g(r) is isotropic in nature with a rapid damping of the oscillations, while the orientational correlation function shows exponential decay; these are signatures of a fluid or liquid phase. At the intermediate temperature of 310 K the g(r) is isotropic, with oscillations that are longer in range, while the orientational correlation function decays algebraically, i.e., as *l/r*; this is a clear signature of a hexatic phase with short range positional order, but quasi-long range orientational order. At the lowest temperature of 280 K the $g(r)$ is anisotropic and typical of a somewhat disordered crystal, while the orientational correlation function shows long range order, again indicating a crystal phase. Examination of snapshots showed that the low temperature phase had a 2-D hexagonal crystal structure.

3.6   Discussion and conclusions

Both the experimental and simulation results show that aniline confined within ACF melts at a temperature higher than the bulk value of 267 K. The experiment gave a melting temperature for the confined system of 298 K, while simulation gave 296 K, an elevation due to confinement of 31 and 29 K, respectively. Such an increase is expected based on our knowledge of the global freezing behavior [7] in view of the high value of $\alpha = 1.2$ for this system, and is consistent with other experimental results for confinement in ACF [5–9].

The experiments show two transitions for the confined system, one at 298 and the second at 324 K. Analysis of the dielectric relaxation times for the three phases shows that for temperatures below 298 they are character-istic of a crystal aniline phase, while above 324 K they are characteristic of liquid phases. For the intermediate phase, between 298 and 324 K, the dielec-tric relaxation times are of the order 10^{-5}s, which is of the order found for hexatic phases. The simulations also show two transitions, at 296 and 336 K, respectively. Analysis of the positional and orientational pair correlation functions shows that the intermediate phase is a hexatic phase, and is the stable phase between 296 and 336 K; thus the lower transition temperature corresponds to melting of the hexagonal crystal to a hexatic phase, while the upper transition temperature is for a hexatic to liquid transition.

The molecular models used in the simulations (slit pore, smooth walls, simple Lennard–Jones potentials) are crude. Nevertheless we believe they capture the physics involved in these transitions. There is good qualitative agreement between the experiment and simulation, and even fairly good quantitative agreement. The results suggest that confinement within narrow slit pores having strongly adsorbing walls, thus enforcing strong layering of the adsorbate, promotes the stability of a hexatic phase, so that it can be observed for simple adsorbate molecules over a rather wide temperature range, 26 K in the experimental system. This is in contrast to previous studies, which have usually been for thin films of liquid crystal phases, where the hexatic phase is only stable over a narrow temperature range. Thus, such confined systems seem promising for further study of hexatic phases.

Acknowledgments

This work was supported by grants from the National Science Foundation (grant no. CTS-9908535) and KBN. Supercomputer time was provided under a NSF/NRAC grant (MCA93S011).

References

1. Miyahara, M. and Gubbins, K.E. (1997) "Freezing/melting phenomena for Lennard–Jones methane in slit pores: a Monte Carlo study," *J. Chem. Phys.* 106, 2865–2880.
2. Radhakrishnan, R. and Gubbins, K.E. (1999) "Free energy studies of freezing in slit pores: an order-parameter approach using Monte Carlo simulation," *Mol. Phys.* 96, 1249–1267.
3. Śliwinska-Bartkowiak, M., Gras, J., Sikorski, R., Radhakrishnan, R., Gelb, L.D., and Gubbins, K.E. (1999) "Phase transitions in pores: experimental and simulation studies of melting and freezing," *Langmuir* 15, 6060–6069.
4. Dominguez, H., Allen, M.P., and Evans, R. (1998) "Monte Carlo studies of the freezing and condensation transitions of confined fluids," *Mol. Phys.* 96, 209–229.
5. Kaneko, K., Watanabe, A., Iiyama, T., Radhakrishnan, R., and Gubbins, K.E. (1999) "A remarkable elevation of freezing temperature of CCl_4 in graphitic micropores," *J. Phys. Chem. B* 103, 7061–7063.
6. Radhakrishnan, R., Gubbins, K.E., Watanabe, A., and Kaneko, K. (1999) "Freezing of simple fluids in microporous activated carbon fibers: comparison of simulation and experiment," *J. Chem. Phys.* III, 9058–9067.
7. Radhakrishnan, R., Gubbins, K.E., and Śliwinska-Bartkowiak, M. (2000) "Effect of the fluid–wall interaction on freezing of confined fluids: towards the development of a global phase diagram," *J. Chem. Phys.* 112, 11048–11057.
8. Gelb, L.D., Gubbins, K.E., Radhakrishnan, R., and Śliwinska-Bartkowiak, M. (1999) "Phase separation in confined systems," *Rep. Prog. Phys.* 62, 1573–1659.
9. Śliwinska-Bartkowiak, M., Dudziak, G., Sikorski, R., Gras, R., Gubbins, K.E., and Radhakrishnan, R. (2001) "Dielectric studies of freezing behavior in porous materials: water and methanol in activated carbon fibers," *Phys. Chem. Chem. Phys.* 3, 1179–1184.
10. Radhakrishnan, R., Śliwinska-Bartkowiak, M., and Gubbins, K.E. (2001) "Global phase diagrams for freezing in porous media," *J. Chem. Phys.* submitted (2000).
11. Masuda, H. and Fukuda, K. (1995) "Ordered metal nanohole arrays made by a two-step replication of honeycomb structures of anodic alumina," *Science* 268, 1466–1468.
12. Harris, P.J.F. (1999) Carbon Nanotubes and Related Structures. New Materials for the Twenty-first Century (Cambridge University Press, New York).
13. Zhakidov, A.A., Baughman, R.H., Iqbal, Z., Cui, C.X., et al. (1998) "Carbon structures with three-dimensional periodicity at optical wavelength," *Science* 282, 897–901.
14. Zhang, Z.B., Gekhtman, D., Dresselhaus, M.S., and Ying, J.Y. (1999) "Processing and characterization of single-crystalline ultrafine bismuth nanowires," *Chem. Mater.* II, 1659–1665.
15. Heremans, J., Thrush, C.M., Lin, Y.M., Cronin, S., et al. (2000) "Bismuth nanowire arrays: synthesis and galvanometric properties," *Phys. Rev. B* 61, 2921–2930.
16. Liu, K., Chien, C.L., Searson, P.C., and Kui, Y.Z. (1998) "Structural and magneto-transport properties of electrodeposited bismuth nanowires," *Appl. Phys. Lett.* 73, 1436–1438.
17. Hill, N., Vaughan, W.E., Price. A.H., and Davies, M. (1970) "Dielectric properties and molecular behaviour," Sugden, T.M., ed, (Van Nostrand Reinhold Co., New York).

18. Ahadov, A.U. (1975) Dielektricieskoje svoistva tchistih zhidkosti (Izdaitelstwo Standardow, Moskva).
19. Szurkowski, B., Hilczer, T., and Śliwinska-Bartkowiak, M. (1993), *Berichte Bunsenges. Phys. Chem.* 97, 731.
20. Śliwinska-Bartkowiak, M., Dudziak, G., Sikorski, R., Gras, R. Radhakrishnan, R., and Gubbins, K.E. (2001) "Melting/freezing behavior of a fluid confined in porous glasses and MCM-41: dielectric spectroscopy and molecular simulation," *J. Chem. Phys.* 114, 950–962.
21. Steele, W.A. (1973), *Surf. Sci.* 36, 317.
22. Kofke, D. (1993), *J. Chem. Phys.* 98, 4149.
23. Agrawal, R. and Kofice, D. (1995). *Mol. Phys.* 85, 43.
24. Hirschfelder, J.O., Curtiss, C.F., and Bird, R.B. (1954) Molecular Theory of Gases and Liquids (Wiley, New York).
25. Nelson, D.R. and Halperin, B.I. (1979), *Phys. Rev. B* 19, 2457.
26. Kosterlitz, J.M. and Thouless, D.J. (1973), *J. Phys. C* 6, 1181.
27. Brock, J.D., Birgenau, R.J., Lister, J.D., and Aharony, A. (1989), *Phys. Today* July, 52.
28. Zangi, R. and Rice, S.A. (1998), *Phys. Rev. E* 58, 7529.

chapter four

Synthesis and characterization of templated mesoporous materials using molecular simulation

F. R. Siperstein

K. E. Gubbins*

North Carolina State University

Contents

4.1   Introduction

Templated mesoporous materials have attracted a lot of attention from the scientific community since their introduction in 1992 by researchers at Mobil [2] although synthesis of similar materials had been reported earlier in the

*Corresponding author.
Reprint from *Molecular Simulation*, 27: 5–6, 2001. http://www.tandf.co.uk

literature dating back to a patent filed in 1969 [3] which was later shown to yield a material with the same properties as MCM-41 [4] and the synthesis of FSM-16 in 1990 [5]. Several synthesis procedures have been proposed since then, using a variety of surfactant-inorganic pairs. Although some syntheses have been performed using a *true liquid crystal templating* technique [6] where the initial surfactant solution is at a high enough concentration to form liquid crystal phases, most syntheses start with a dilute surfactant solution [7]. In this case, the final structure of the porous solid does not resemble the micellar structure in the surfactant solution used for the synthesis. The prediction of the final structure of the porous materials is complicated because the physics underlying these syntheses is not well understood, mainly due to the over-lapping self-assembly and inorganic polymerization processes.

The key to predicting structures of templated mesoporous materials is to understand how a dilute solution with spherical micelles becomes a liquid crystal phase when silica is added to the system. The detailed sim-ulation of micelle formation alone is a challenging problem for today's computer power. Simulations using detailed atomistic potentials have, for the most part, focused on the evolution of prearranged structures, but are usually unable to span real times that are long enough to observe formation and destruction of micelles [8]. Coarse-grained approaches, either on- or off-lattice, are normally used to study the formation of micelles. Although off-lattice simulations are typically more versatile and realistic, lattice sim-ulations allow larger systems to be studied, and can be made more realistic through lattice discretization [9].

Lattice Monte Carlo simulations have been used to study micellization processes and to determine binary and ternary phase diagrams containing spherical micelles, hexagonal, lamellar and cubic structures. Generally the model surfactant molecules are made up of m hydrophilic head groups H and n hydrophobic tail groups T, i.e. $H_m T_n$, and are distributed across lattice sites with one group per site. Solvent molecules, S, occupy single sites, and oil molecules if present occupy one or several sites [10–17]. Recently lattice Monte Carlo simulations have been used to study evaporation driven self-assembly to describe dip-coating synthesis of templated materials [18].

In this work, we show that synthesis of templated materials in bulk solution can be interpreted with an equilibrium triangular diagram for sur-factant, solvent and silica where different liquid crystal phases can be located. The liquid crystal-like behavior of silica-surfactant phases has been observed experimentally under no polymerization conditions [19]. The equi-librium diagram is calculated using a lattice Monte Carlo approach, speci-fying the appropriate interaction parameters to represent each component. Two extreme cases are studied, one where the three binaries are completely miscible and another where the solvent and the inorganic oxide (silica) are immiscible. In the former case, a favorable interaction between the silica and the surfactant head produces an immiscibility gap inside the ternary dia-gram, while in the latter case a larger region of phase separation occurs due to the immiscibility between the solvent and the oxide. The shape and location of the immiscibility gap determines the different liquid crystal phases that can be formed.

Model porous materials are obtained by assuming that the silica polymerization is sufficiently fast that no modification of the liquid crystal structure occurs. Silica polymerization is followed by removal of the surfactant chains. Adsorption isotherms and heats of adsorption of argon on a model material with cylindrical pores are calculated using grand canonical Monte Carlo simulations. Heats of adsorption, calculated from fluctuations in the energy and number of molecules [1] during the GCMC simulations, decrease with coverage for loadings below one statistical monolayer, in agreement with experimentally measured heats of adsorption of argon and krypton on MCM-41 [20,21].

4.2 Simulation technique

4.2.1 Lattice Monte Carlo

We used Larson's lattice model [10] with a fully occupied cubic lattice in which a site interacts equally with all 26 sites that lie within one lattice spacing in directions $(1,0,0)$, $(1,1,0)$ and $(1,1,1)$. Each segment of the surfactant ($H_m T_n$), solvent (S) or silica (inorganic oxide, I) occupies a single point on the lattice. $H_4 T_4$ was used as a model surfactant molecule, which consists of a sequence of four hydrophilic head segments H, and four hydrophobic tail segments T. A site on the surfactant chain can be connected to any of its $z = 26$ nearest-neighbors or diagonally nearest-neighbors.

Each molecular unit (H, T, S and I) is characterized by an interaction energy ε_{ab}(a, b = H, T, S, I). The net energy change associated with any configuration rearrangement depends on a set of interchange energies ω_{ab}:

$$\omega_{ab} = \varepsilon_{ab} - \frac{1}{2}(\varepsilon_{aa} + \varepsilon_{bb}) \quad \text{with a} \neq \text{b} \qquad (4.1)$$

and not on the individual interaction energies ε_{ab}. The surfactant-solvent interaction parameters are the same as those used by other researchers [14–16]: $\omega_{HT}/k_B T = \omega_{ST}/k_B T = 0.153846$ and $\omega_{HS} = 0$. A strong inorganic-head attraction to mimic the strong affinity between silica and surfactant heads was specified as $\omega_{IH}/k_B T = -0.307692$ and $\omega_{TT} = 0.153846$. Two extreme cases were studied, one where the inorganic component and the solvent are completely miscible ($\omega_{IS} = 0$) and another where they are immiscible ($\omega_{IS}/k_B T = 0.153846$). The reduced temperature is defined using the head–tail DA6"\char"32}tail interchange energy by $T^* = k_B T/\omega_{HT}$.

All Monte Carlo simulations were performed in the canonical ensemble (NVT) with periodic boundary conditions. Reputation and "kink" -like moves were considered in addition to chain regrowth using the configurational bias method [22]. Ternary liquid–liquid equilibria were calculated using a direct interfacial approach. One dimension of the simulation box was increased with respect to the other two by a factor of eight to make the formation of planar interfaces preferable to curved interfaces [15]. The box size used in the simulation was $40 \times 40 \times 320$. Bulk coexisting densities were estimated from ensemble averages of the system densities away from the interfaces. The reduced temperature was chosen to be the same as that used by Larson in his work ($T^* = 6.5$). Up to 125,000 cycles were needed for the system to equilibrate, where each cycle consists of $40 \times 40 \times 320$ configurations.

Different density profiles were used as the initial configuration for the ternary liquid–liquid equilibrium calculations, as a check that the system had reached equilibrium. The starting configuration was obtained as follows. A high surfactant concentration slab was obtained by equilibrating at infinite temperature a $40 \times 40 \times 40$ or $40 \times 40 \times 80$ box with periodic boundary conditions in the x and y directions and had walls in the z direction. Typically, the high-surfactant concentration box contained 60% in volume of surfactant while the rest was solvent. After equilibration, the box with the high surfactant concentration was placed in the $40 \times 40 \times 320$ box, resulting in a slab having 60% surfactant. Silica and solvent units were distributed randomly over the whole $40 \times 40 \times 320$ box not allowing for overlaps with the previously arranged surfactant chains. This selection of initial configuration favored the formation of only two interfaces (Figure 4.1). The final surfactant concentration in the high surfactant concentration slab varied between 40 and 80% in volume when phase separation was observed. Formation of spherical micelles was observed in the absence of silica.

Some simulations were carried out starting from a completely homogeneous box, but the equilibration time was considerably larger and the formation of more than two interfaces was observed. Nevertheless, the compositions far from the interface were the same for different initial configurations.

Figure 4.1 Initial configuration (top) and snapshot after 15×10^9 configurations (bottom) at $T^* = 6.5$ for the system with complete miscibility between solvent and silica. Surfactant heads are in yellow, surfactant tails are in red and silica units are in gray. The system is converging into a lamellar phase.

The materials obtained from the direct interfacial simulation do not present perfectly flat interfaces, and are not convenient to use for adsorption measurements because the interface curvature generates unrealistic large pores when periodic boundary conditions are used in the x, y and z directions. Removing the interface curvature does not yield a proper periodic material. Therefore, some simulations were carried out at the bulk density of the high-surfactant high-silica phase obtained from the liquid–liquid equilibrium calculation to obtain model materials without having the problems due to the interfaces. These simulations were performed in a cubic box (between 20^3 and 80^3) with periodic boundary conditions. A typical run consisted of $2 - 7 \times 10^8$ configurations for the system to equilibrate.

After the system reached equilibrium it was assumed that connected silica units polymerize without modification of the mesostructure, and the surfactant and unconnected silica units were removed from the system.

4.2.2 Material characterization

Adsorption isotherms and heats of adsorption of argon in a prototype material with cylindrical pores obtained from the mimetic synthesis were calculated using grand canonical Monte Carlo simulations. The material was obtained at a surfactant concentration of 55% and silica concentration of 35%, with the remainder being solvent. The structure of the material as obtained from the lattice simulation is shown in Figure 4.2. The material shows cylindrical pores with diameter of eight lattice segments.

The positions of the silica segments obtained from the lattice simulation were scaled to obtain a material with a pore diameter of 4 nm. Thus, the distance between each lattice point was assumed to be 0.5 nm. A sphere of 0.5 nm is considerably larger than the oxygen diameter in silica materials (0.27 nm); therefore, it was assumed that each silica sphere corresponded to a collection of silica units. The distance between connected silica spheres in the lattice can be between 1 and 1.73 ($\sqrt{3}$) times the separation between lattice points. Therefore, each silica sphere was assumed to have a hard core center of 0.72 nm to avoid the formation of micropores between each silica sphere. Each sphere has a density of 2.7 g/cm^3 [23] and the interactions between these spheres and adsorbed fluids were modeled following the work of Kaminsky and Monson [24], where the solid–fluid potential is given by:

$$u_{sf}(r) = \frac{16}{3}\pi\epsilon_{sf}\rho_s R^3 \left(\frac{(r^6 + (21/5)r^4 R^2 + 3r^2 R^4 + (1/3)R^6)\sigma_{sf}^{12}}{(r^2 - R^2)^9} - \frac{\sigma_{sf}^6}{(r^2 - R^2)^3} \right) \quad (4.2)$$

where R is the hard core radius of the solid sphere and ρ_s its density; ε_{sf} and σ_{sf} are the Lennard–Jones interaction parameters between a fluid molecule and an oxygen atom in a siliceous material that were taken from argon–oxygen interaction parameters in silicalite [25]. Parameters used for the simulation

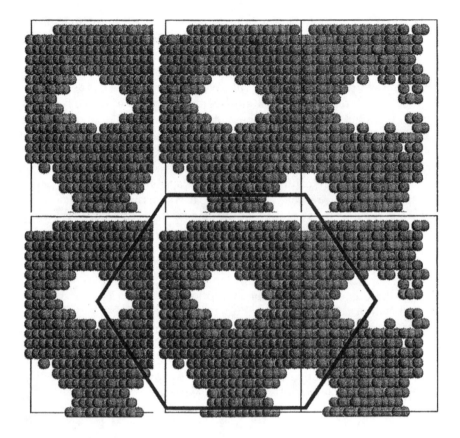

Figure 4.2 Model silica structure showing the hexagonal arrangement of independent cylindrical pores. $2 \times 2 \times 1$ simulation boxes are shown.

are summarized in Table 4.1. The solid sphere radius, R, was taken as 3.25 Å to allow some overlap between the silica spheres.

Simulations were performed using periodic boundary conditions. The system was allowed to equilibrate during the first $1 \times 10^5 - 1 \times 10^6$ cycles of each simulation point. Statistics were taken over the following one million cycles. Each cycle corresponded to a displacement and a creation or deletion. Each simulation was divided into ten blocks. Partial averages in each block were used to calculate standard deviations in total amount adsorbed and energy of the system. Simulations were carried out at 77 K for pressures up to 0.5 bar.

Table 4.1 Constants of Lennard–Jones12–6 potential [25]

	σ(nm)	ε/k (K)
Fluid–fluid	0.3405	119.8
Fluid–solid	0.3335	93.0

Heats of adsorption were calculated as [1]:

$$q_{st} = -\frac{f(N,U)}{f(N,N) - \bar{N}^G} + kT \tag{4.3}$$

where the fluctuations are defined as $f(X, Y) = \langle XY \rangle - \langle X \rangle \langle Y \rangle$ and $\bar{N}^G$ is the number of molecules at the bulk gas density in the same volume as the system. In our case, $\bar{N}^G$ was ignored because it is negligible compared with $f(N,N)$.

4.3   Results

4.3.1   Synthesis of silica materials

The phase diagram obtained for solvent-H_4T_4 was in good agreement with the one reported by Larson [14]. Addition of a silica source to a surfactant solution results in a phase separation where a surfactant-rich silica-rich phase is in equilibrium with a surfactant-poor silica-poor phase [9]. The surfactant rich phase presents liquid crystal type behavior.

Phase separation in a ternary system where two binaries are completely miscible can be achieved either by an effective attraction between two components strong enough to form a two-phase region, or by an effective repulsion between molecules of two different components that is sufficiently strong to induce the phase separation. The former process is known as associative and the latter as segregative phase separation [26]. These types of phase separation have been observed in systems containing a polyelectrolyte (hyaluronate) and a cationic surfactant (alkyltrimethyammonium-bromide) [27].

Ternary diagrams for associative and segregative phase separations are shown in Figure 4.3. The formation of a surfactant-rich phase is observed in both cases. This phase can adopt different liquid–crystal type structures, such as lamellar, perforated lamella, and hexagonal The formation of a bicontinuous phase was observed, but not of a cubic phase. The formation of a cubic phase is not observed when the size of the simulation box does not correspond to an integer multiple of the unit cell parameters of the cubic structure. More detailed studies are needed in the borders between lamellar and hexagonal phases to observe the formation of cubic phases.

The associative phase diagram is similar to the behavior observed for the ternary system water–sodium hyaluronate (NaHy)-alkyltrimethylammonium bromide (CTAB) in the absence of salt [27]. This behavior was expected, since at the synthesis conditions the silica source is a highly charged oligomer. An important difference between the solvent–silica–CTAB and water–NaHy–CTAB systems is that for the latter the high-surfactant concentration phase contains no more than 30% of surfactant, which probably is not enough to observe the formation of surfactant liquid–crystal phases.

Some general trends observed experimentally were found in our mimetic synthesis. For example, variation of the surfactant/silica ratio result in the formation of different mesophases [28]. Hexagonal phases are observed for

Adsorption and Transport at the Nanoscale

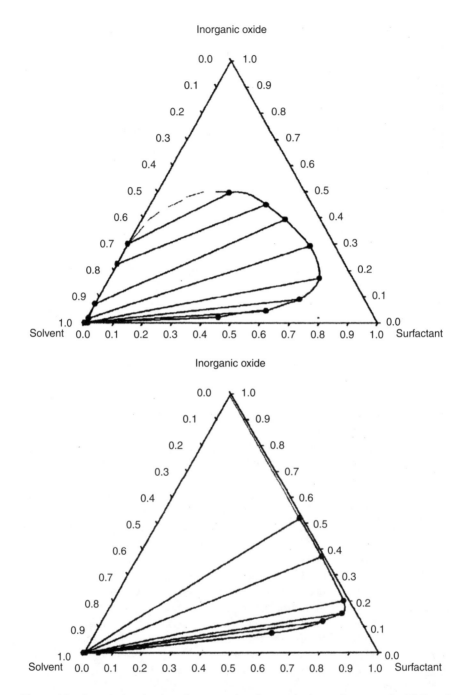

Figure 4.3 Associative (top) and segregative (bottom) phase diagrams for H_4T_4–solvent–silica at $T^* = 6.5$.

low surfactant/silica ratios (*ca.* 0.6) and lamellar phases are observed for higher surfactant/silica ratios (*ca.* 1.3). We observe a lamellar phase for surfactant/silica ratios of 1.0 and hexagonal phases for surfactant/silica ratios of 0.13–0.20. These ratios are practically independent of whether the mixture is associative or segregative. The exact boundaries for these phases have not been calculated. At high surfactant/silica ratios (*ca.* 2) perforated lamella are observed and for surfactant/silica ratios of 10 it is observed that the silica is deposited between adjacent micelles.

No cubic phases were found, which experimentally are observed at a surfactant/silica ratio of 1.0. The cubic octamer, observed experimentally at a surfactant/silica ratio of 1.9, is not observed, mainly due to the lack of structure of the silica in our simulations and to the different number of possible interactions between one surfactant chain and silica units. Experimental evidence shows that one silica unit interacts with the ammonium group of one surfactant chain, while in our simulations the limit is specified by the connectivity of the lattice. Another important difference is that a commonly used source of silica for the synthesis of templated materials is TEOS (tetraethoxiorthosilane), which produces ethanol upon polymerization. The addition of an alcohol to a surfactant system changes the solubility of the surfactant and the structure of the micelles and liquid crystal phases, which is a phenomena not accounted for in our simulations.

4.3.2 Material characterization

Simulation studies of gas adsorption on MCM-41 type materials using an idealized pore geometry show that surface energetic heterogeneity needs to be considered in order to describe correctly low coverage nitrogen adsorption isotherms [23]. Heats of adsorption of simple fluids (argon [20] and krypton [21]) on MCM-41 decrease by approximately 3 kJ/mol on going from zero coverage to half saturation, confirming the adsorbent heterogeneity.

Heats of adsorption on homogeneous cylinders calculated using non-local DFT [29] increase with coverage as a result of adsorbate–adsorbate interactions. When surface heterogeneity is present, as in corrugated surfaces, heats of adsorption decrease with coverage [30].

Heats of adsorption calculated in our model material decrease with coverage from about 13 down to 6 kJ/mol (Figure 4.4). Our model presents less high-energy sites than what would be expected from experimental measurements, consequently, the decrease in the heats of adsorption in our modeled material extends only to approximately 20% of saturation whereas in real materials it extends to 40% of saturation.

An argon adsorption isotherm calculated on our model material is compared with experimental isotherms in Figure 4.5. The calculated isotherm has qualitatively the same shape as the experimental one, and differences between experimental and simulated isotherms are consistent with the differences observed in the heats of adsorption. The uptake at low pressures is

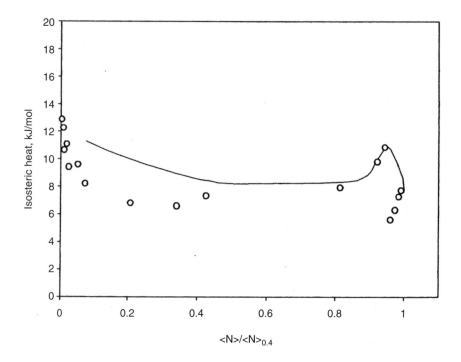

Figure 4.4 Isosteric heats of adsorption for argon on MCM-41 type material at 77 K. Symbols are simulation results and solid line shows experimental results [20]. The simulated amount adsorbed is normalized by the amount adsorbed at $f/f^0 = 0.4$ and experimental results at $P/P^0 = 0.8$.

less than what is observed experimentally, which is consistent with a smaller number of high-energy sites compared with real materials.

Snapshots of the simulation at low and high relative pressure are shown in Figure 4.6. The influence of the lattice used for the synthesis is evident from the solid structure. At low relative pressures a cylindrical monolayer is formed and at high relative pressures the complete pore is filled with argon.

4.4 Conclusions

We have developed a methodology to determine silica porous structures following a typical templating material synthesis in bulk solution. The range of structures obtained is in qualitative agreement with experimental observations: hexagonal phases are observed at low surfactant/silica ratios and lamellar phases at high surfactant/silica ratios.

Grand canonical Monte Carlo simulations of argon adsorption were used to characterize the materials obtained with the mimetic simulation. Adsorption properties in the materials modeled are comparable with experimental measurements and they indicate that the adsorbent has less high-energy sites

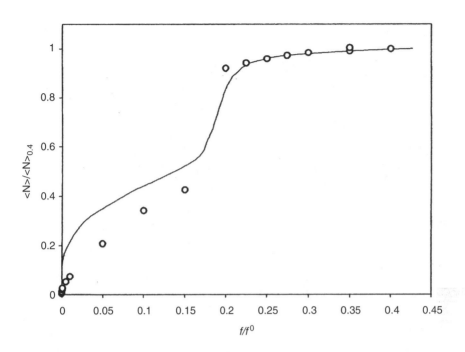

Figure 4.5 Argon adsorption isotherms at 77 K for argon on MCM-41. Open symbols are simulation results and the solid line is an experimental isotherm taken from Ref. [20].

than real materials. A more detailed description of the silica spheres may account for a broader surface heterogeneity.

The choice of surfactant produces a material with walls that are considerably thicker than in MCM-41 type materials. The wall thickness depends

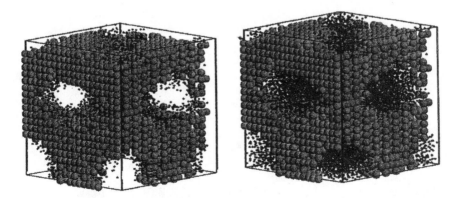

Figure 4.6 Snapshot of adsorption simulation at $f/f^0 = 0.1$ (left) and $f/f^0 = 0.2$ (right). Solid silica structure is in gray and argon is in blue. Argon atoms are shown to a reduced scale for better visualization.

on the length of the hydrophilic section of the surfactant; thus using a surfactant with a smaller head to tail ratio will yield a more realistic material.

The structure of silica materials is not well reproduced by cubic lattices because the tetrahedral arrangement between silica units cannot be reproduced. Future work will be concentrated on using a more realistic description of the silica for the synthesis of templated materials.

Acknowledgments

We thank the Department of Energy for support of this research under grant no. DE-FG02 98ER14847. FRS thanks Martin Lísal for many useful discussions.

References

1. Nicholson, D. and Parsonage, N.G. (1982) Computer Simulation and the Statistical Mechanics of Adsorption (Academic Press, London), p 97.
2. Beck, J.S., Vartuli, J.C., Roth, W.J., Leonowicz, M.E., Kresge, C.T., Schmitt, K.D., Chu, C.T.-W., Olson, D.H., Sheppard, E.W., McCullen, S.B., Higgins, J.B. and Schlenker, J.L. (1992) "A new family of mesoporous molecular sieves prepared with liquid crystal templates," *J. Am. Chem. Soc.* 114, 10834.
3. Chiola, V., Ritsko, J.E. and Vanderpool, C.D. (1971) US Patent 3 556 725.
4. DiRenzo, F., Cambon, F.H. and Dutartre, R. (1997) "A 28-year-old synthesis of micelle templated mesoporous silica," *Micropor. Mater.* 10, 283.
5. Yanagisawa, T., Shimizu, T., Kuroda, K. and Kato, C. (1990) "The preparation of alkyltrimethylammonium-kanemite complexes and their conversion to microporous materials," *B. Chem. Soc. Jpn.* 63, 988.
6. Attard, G.S., Glyde, J.C. and Goltner, C.G. (1995) "Liquid-crystalline phases as templates for the synthesis of mesoporous silica," *Nature* 378, 366.
7. Ciesla, U. and Schüth, F. (1999) "Ordered mesoporous materials," *Micropor. Mesopor. Mater.* 27, 131.
8. Shelley, J.C. and Shelley, M.Y. (2000) "Computer simulation of surfactant solutions," *Curr. Opin. Colloid Interf. Sci.* 5, 101.
9. Panagiotopoulos, A.Z. (2000) "On the equivalence of continuum and lattice models for fluids," *J. Chem. Phys.* 112, 7132.
10. Larson, R.G., Scriven, L.E. and Davis, H.T. (1985) "Monte Carlo simulation of model amphiphile-oil-water system," *J. Chem. Phys.* 83, 2411.
11. Larson, R.G. (1992) "Monte Carlo simulation of microstructural transitions in surfactant systems," *J. Chem. Phys.* 96, 7904.
12. Larson, R.G. (1989) "Self assembly of surfactant liquid crystalline phases by Monte Carlo simulation," *J. Chem. Phys.* 91, 2479.
13. Talsania, S.K., Wang, Y., Rajagopalan, R. and Mohanty, K.K. (1997) "Monte Carlo simulations for micellar encapsulation," *J. Colloid Interf. Sci.* 190, 92.
14. Larson, R.G. (1996) "Monte Carlo simulations of the phase behavior of surfactant solutions," *J. Phys. II France* 6, 1441.
15. Mackie, A.D., Onur, K. and Panagiotopoulos, A.Z. (1996) "Phase equilibria of a lattice model for an oil–water–amphiphile mixture," *J. Chem. Phys.* 104, 3718.
16. Mackie, A.D., Panagiotopoulos, A.Z. and Szleifer, I. (1997) "Aggregation behavior of a lattice model for amphiphiles," *Langmuir* 13, 5022.

17. Floriano, M.A., Onur, K. and Panagiotopoulos, A.Z. (1999) "Micellization in model surfactant systems," *Langmuir* 15, 3143.

18. Malanoski, A.P. and Van Swol, F. (2000) "Lattice models for adsorption and transport in self-assembled nano-structures," *AIChE Annual Meeting*.

19. Firouzi, A, Atef, F., Oertii, A.G., Stucky, G.D. and Chmelka, B.F. (1997) "Alkaline lyotropic silicate–surfactant liquid crystals," *J. Am. Chem. Soc.* 119, 3596.

20. Neimark, A.V., Ravikovitch, P.I., Grün, M., Schüth, F. and Unger, K.K. (1998) "Pore Size analysis of MCM-41 type adsorbents by means of nitrogen and argon adsorption," *J. Colloid Interf. Sci.* 207, 159.

21. Olivier, J.P. (2000) "Thermodynamic properties of confined fluids. I. Experimental measurements of krypton adsorbed by mesoporous silica from 80 K to 130 K," in: *Proceedings of the Second Pacific Basin Conference on Adsorption Science and Technology* Do, D.D., ed, (World Scientific, Singapore), pp 472–476.

22. Frenkel, D. and Smit, B. (1996) Understanding Molecular Simulation (Academic Press, San Diego).

23. Maddox, M.W., Olivier, J.P. and Gubbins, K.E. (1997) "Characterization of MCM-41 using molecular simulation: heterogeneity effects," *Langmuir* 13, 1737.

24. Kaminsky, R.D. and Monson, P.A. (1991) "The influence of adsorbent microstructure upon adsorption equilibria: investigations of a model system," *J. Chem. Phys.* 95, 2936.

25. Talu, O. and Myers, A.L. (1998) "Force constants for adsorption of helium and argon in high-silica zeolites," Fundamentals of Adsorption 6, Meunier, F., ed, (Elsevier, Paris) pp 861–866.

26. Picullel, L. and Lindman, B. (1992) "Association and segregation in aqueous polymer/polymer, polymer/surfactant and surfactant/surfactant mixtures: similarities and differences," *Adv. Colloid Interf. Sci.* 41, 149.

27. Thalberg, K., Lindman, B. and Karlstrom, G. (1990) "Phase diagram of a system of cationic surfactant and anionic polyelectrolyte—tetradecyltrimethylammonium bromide–hyaluronan–water," *J. Phys. Chem.* 94, 4289.

28. Vartuli, J.C., Schmitt, K.D., Kresge, C.T., Roth, W.J., Leonowicz, M.E., McCullen, S.B., Hellring, S.D., Beck, J.S., Schlenker, J.L., Olson, D.H. and Sheppard, E.W. (1994) "Effect of surfactant/silica molar ratios on the formation of mesoporous molecular sieves: inorganic mimicry of surfactant liquid–crystal phases and mechanistic implications," *Chem. Mater.* 6, 2317.

29. Balbuena, P.B. and Gubbins, K.E. (1994) "The effect of pore geometry on adsorption behavior," Characterization of Porous Solids III, Studies in Surface Science and Catalysis, Rouguerol, J., Rodriguez-Reinoso, F., Sing, K.S.W. and Unger, K. POK., eds, (Elsevier, Amsterdam) 87, pp 41–50.

30. Steele, W. and Bojan, M.J. (1997) "Computer simulation study of sorption in cylindrical pores with varying pore wall heterogeneity," Characterization of Porous Solids IV (The Roy Chemical Society, Cambridge), pp 49–56.

chapter five

Adsorption/condensation of xenon in mesopores having a microporous texture or a surface roughness

R. J.-M. Pellenq
Centre de Recherche sur les Mécanismes de la Croissance Cristalline

B. Coasne
Groupe de Physique des Solides

P. E. Levitz
Laboratoire de Physique de la Matière Condensée

Contents

Reprint from *Molecular Simulation*, 27: 5–6, 2001. http://www.tandf.co.uk

5.1   Introduction

It is known from theoretical and simulation studies on simple pore geometry (slits and cylinders) that confinement strongly influences the thermodynamics of confined fluid [1]. However, the effect of matrix disorder in terms of pore morphology (pore size and shape), topology (the way the pores distribute and connect in space) and surface roughness on the thermodynamics of confined molecular fluids still remains to be clarified.

Real silica mesoporous materials exhibit different types of disorder. For regular cylindrical pores, MCM-41, the exact nature (roughness or microporosity) of the pore surface texture is still under investigation [2]. In the case of another mesoporous silica material, SBA-15, cylindrical pores are known to be connected by intra-wall microporous channels (i.e., pores of a few molecular diameter large) [3]. Controlled porous glasses (CPG) constitute another class of silica materials among which Vycor is. Although being a disordered material having a rough inner interface [4], Vycor is also known to exhibit no real microporosity. However CPG's can exhibit microporosity depending on synthesis conditions and chemical and heat treatments. Despite many characterization studies, the distinction between surface roughness and microporosity remains unclear: the core matrix of a mesoporous solid limited by a rough interface with a typical characteristic length on the order of a few nanometers can be considered as containing a microporous texture.

The questions addressed in this chapter are thus the following: (*i*) what is the effect of the surface roughness or microporosity on adsorption/ condensation phenomena of fluids confined in real mesoporous solids? (*ii*) Do surface roughness and microporosity lead to different effects on adsorption/ condensation phenomena in the mesoporous regions? In other words, does gas adsorption enable to unambiguously distinguish surface roughness from microporosity? In order to get some insights on those questions, we have simulated by means of Grand Canonical Monte Carlo technique (GCMC), Xe adsorption at 195 K in mesoporous solids having either a microporous texture or a nanometric surface roughness. This study is carried out for ordered (MCM-41 type) and disordered (controlled porous glass) atomistic silica mesopores. Results are compared with those obtained in the case of a silica mesoporous solid having smooth cylindrical pores.

We have also calculated the reference adsorption isotherm and isosteric heat curves for Xe in silicalite at the same temperature (silicalite being a pure siliceous zeolite).

This chapter is organized as follows. The second section presents the numerical procedure used to prepare atomistic silica pores of various morphologies and topologies and the GCMC technique used to simulate adsorption/desorption processes. The third section compares results for xenon adsorption and condensation at 195 K in mesoporous solids having either a microporous texture or a surface roughness.

5.2 Computational details

5.2.1 Generating porous solids

5.2.1.1 Silicalite

The silicalite-1 crystal structure has a porous network consisting of straight channels crossing so-called zig-zag channels; both types having their diameter around 5 Å. This microporous crystal belongs to the *Pnma* symmetry group. The crystallographic cell contains 288 atoms ($Si_{96}O_{192}$) with lattice parameters a = 20.07 Å, b = 19.92 Å and c = 13.42 Å [5]. The GCMC simulations in our study were performed at 195 K in a periodic simulation box containing three silicalite unit cells stacked along the c-direction.

5.2.1.2 Rough/smooth pore

All other porous matrices used in our simulations were prepared from a cubic non-porous siliceous solid (cristoballite). We cut out portions of this initial volume in order to obtain different porous media (from a single regular cyndrical pore to a disordered porous matrix made of interconnected pores of a complex morphology). In order to model the pore inner surface in a realistic way, we first remove all silicon atoms that are in an incomplete tetrahedral environment. At a second step, we remove all non-bonded oxygen atoms (two dangling bonds). This procedure ensures that (*i*) all silicon atoms have no dangling bonds and (*ii*) oxygen atoms have at least one bond with a Si atom. Finally, the electroneutrality of the simulation box is ensured by saturating all oxygen dangling bonds with hydrogen atoms. The latter are placed in the pore void, perpendicularly to the pore surface, at a distance of 1 Å from the closest unsaturated oxygen atom. Regular or irregular cylindrical pores can be easily prepared with this procedure: pore voids are defined by simple mathematical functions. We have prepared a rough cylindrical pore composed of several strips having the same thickness (1.07 nm) but different random diameters. The mean pore size is 4.12 nm and the dispersion along the pore axis is about 1 nm (strip radii are reported in Table 5.1). We have also prepared a smooth atomistic cylindrical pore having a diameter of 4.12 nm. Figure 5.1 shows a transversal view of the rough and smooth 4.12 nm pores.

Table 5.1 Series of strip radii used to prepare
the rough pore shown in Figure 5.1(a)

Strip	Radius (nm)	Strip	Radius (nm)
1	2.00	6	2.34
2	2.50	7	1.76
3	1.58	8	2.22
4	2.15	9	2.03
5	1.63	10	2.42

5.2.1.3 Vycor porous glass

The numerical method described in the previous subsection can be also used to generate complex porous structures. However, the case of disordered porous solids needs an important effort to produce 3D numerical matrices and account for the morphology and the topology of the real material. As far as mesoporous Vycor is concerned, we have used an off-lattice reconstruction algorithm in order to numerically generate the mesoporous region, which has the main morphological and topological properties of real (low-specific surface area -100 m^2/g) Vycor in terms of pore shape. The off-lattice functional represents the Gaussian field associated to the volume autocorrelation function of the studied porous structure [3]. It allows cutting

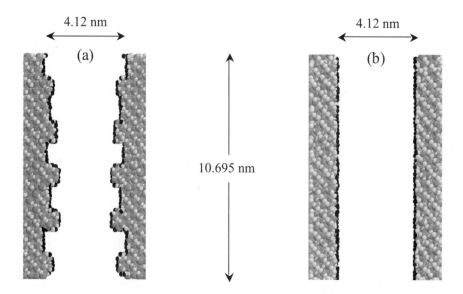

Figure 5.1 (a) Transversal view of a cylindrical silica nanopore with a rough surface. The pore is defined as an assembly of strips (1.07 nm thick) with a random diameter. The average diameter is 4.12 nm and the size dispersion is about 1 nm. (b) Transversal view of a regular cylindrical nanopore with a diameter of 4.12 nm and a length of 10.695 nm. White and gray spheres are respectively oxygen and silicon atoms. Black spheres correspond to hydrogen atoms, which delimitate the pore surface.

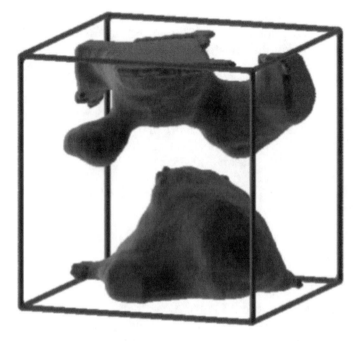

***Figure* 5.2** A numerical reconstruction of 220 m²/g-Vycor. The box size is 10.7 nm. One sees through the silica matrix; the porosity is in grey.

out portions of the initial volume in order to create, in our case, the meso-porosity of the resulting structure (assuming a mesoporosity at $\phi_{meso} = 30\%$ which corresponds to that of many silica porous glasses including Vycor). Note that this approach encompasses a statistical description: it allows generating a set of morphologically and topologically equivalent numerical samples of pseudo-Vycor. An example of a pure mesoporous pseudo-Vycor is given in Figure 5.2. A close inspection of molecular self-diffusion shows that the off-lattice reconstruction procedure reproduces many properties of real Vycor such as tortuosity, and in and out pore two-point correlation functions [6]. In agreement with the experiments, the small angle scattering spectrum of the reconstructed Vycor shows a correlation peak which corresponds to a minimal (pseudo) unit-cell size around 270 Å [4]; this simulation box is too large to be correctly handled in an *atomistic* Monte-Carlo simulation of adsorption in a reasonable amount of CPU time. Hence, we have applied a homothetic reduction with a factor of 2.5 that preserves the mesoporous morphology but reduces the average pore size from 70–90 Å to roughly 30–35 Å [7]. Interestingly enough, the numerical pseudo-Vycor with a pure mesoporosity has a specific area and an average pore size that are close that of the real high specific surface area Vycor (around 4 nm and 220 m²/g respectively) [8]. The homothetically reduced functional is then applied to cut out the porosity from a cube of cristoballite of (106.95)³ Å³ i.e., containing (15 × 15 × 15) unit cells (the resulting structure is shown in Figure 5.2).

5.2.1.4   Mixed (meso/micro) porous Vycor

An atomistic description of a mixed (meso and micro) porosity solid can be obtained by applying the same off-lattice functional described in the previous subsection to a box containing the silicon and oxygen atoms of 5*5*7 unit cells of orthorhombic silicalite [9] (the simulation cell volume is again 106.95^3 Å^3). Our numerical sample thus contains both *micro* and *meso* porous regions. Obviously the resulting value of the total porosity is not any more that of a pure mesoporous glass since the matrix also contains zeolitic microporosity. One can then consider our mixed structure as a defective silicalite crystal having mesoporous cracks. This may not be far from the reality since hysteretic adsorption isotherms (characteristics of mesoporosity as will be seen below) have been found experimentally in the case of nitrogen adsorption at 77 K in silicalite. [10]

5.2.2   The Grand Ensemble Monte-Carlo simulation

5.2.2.1   Intermolecular potentials

In this work, we have used a PN-TrAZ potential function as reported for adsorption of rare gases in silicalite. It is based on the usual partition of the adsorption intermolecular energy which can be written as the sum of a dispersion interaction term with the repulsive short range contribution and an induction term (no electrostatic interaction in the rare gas/surface intermolecular potential function) [11]. The dispersion and induction parts in the Xe/H potential are obtained assuming that hydrogen atoms have a partial charge of 0.5e (q_O = −1e and q_{Si} = −2e respectively) and a polarizability of 0.58 Å^3. Only the Xe/H repulsive contribution is adjusted on the experimental (Vycor) low coverage isosteric heat of adsorption (Qst (0) = 17 kJ/mol) [12]. One may infer that the isosteric heat of adsorption at zero coverage on a purely mesoporous pseudo-Vycor numerical sample should be higher than that measured on the real material due to higher surface curvature induced by the homothetic reduction. In fact, Qst(0) does not depend strongly upon surface curvature for pores larger than 8 Å in size: in the case the Xe/ silicalite system at 121 K (pore diameter 5 Å), Qst(0) = 27.4 kJ/mol [11,13], it decreases to 17.9 kJ/mol in the cavity of NaY zeolite (pore diameter 8 Å [14]. Note that in the last case, Qst(0) is only 1 kJ/mol larger than that in Vycor. Therefore, we can safely consider that the isosteric heat of adsorption at zero coverage in non-microporous numerical pseudo-Vycor samples is that of the real material that has slightly larger pores than its numerical counterpart. In this work, two Xe/Xe Lennard-Jones potentials have been used. In the case of silicalite and the mixed silicalite/Vycor structure, we have used the potential parameters reported by Barker (ε = 281 K and σ = 3.89 Å) which gives a good fit of the "true" two-body Xe/Xe potential [15]. The corresponding bulk saturation pressure is then 65000 Pa at 195 K according to the Lennard-Jones equation of state proposed by Kofke [16]. The reason for this choice is that in the micropores of silicalite, Xe atoms have a very low coordination number (compared to the bulk liquid), hence a "true" two-body is relevant. By contrast, in nanopores where one expects capillary condensation to occur, an appropriate bulk-like potential should be preferred if one aims at describing the energetics of the liquid phase. Thus, in the case of smooth and rough cylindrical pores, we have used the potential parameters reported by Steele for bulk liquid xenon ($\varepsilon = 211$ K and $\sigma = 0.41$ nm); the bulk saturating pressure being 587366 Pa [16]. In the case of the silicalite/Vycor system, we have chosen the Xe-Xe intermolecular potential, which describes the best properties of the fluid confined in the zeolitic regions.

5.2.2.2   *The Grand Canonical Monte-Carlo technique as applied to adsorption in pores*

In the Grand Canonical Ensemble, the independent variables are the chemical potential, the temperature and the volume [17]. At equilibrium, the chemical potential of the adsorbed phase equals that of the bulk phase which constitutes an infinite reservoir of particles at constant temperature. The chemical potential can be related to the temperature and the pressure of the bulk phase according to the equation of state for an ideal gas. The adsorption isotherm can be readily obtained from such a simulation technique by evaluating the ensemble average of the number of adsorbate molecules. Plots of the number of adsorbed molecules and internal energy *versus* the number of Monte-Carlo steps were used to monitor the approach to equilibrium. Acceptance rates for creation or destruction were also followed and should be equal at equilibrium. After equilibrium has been reached, all averages were reset and calculated over several millions of configurations ($3 \cdot 10^5$ Monte-Carlo steps per adsorbed molecules). In order to accelerate GCMC simulation runs, we have used a grid-interpolation procedure in which the simulation box volume is split into a collection of voxells [13]. The Xe/Silica adsorption potential energy is calculated at each corner of each elementary cube. A cut through a grid is presented in Figure 5.3 in the case of the mixed zeolite/Vycor sample: one can see the microporous zeolitic channels. In our GCMC simulations, periodic boundary conditions have been applied in the x, y and z directions to avoid finite size effects. Note that the off-lattice method used to generate the disordered matrices is adapted from its original version to meet this requirement [6].

5.3   Result and discussion

5.3.1   *Effect of the surface roughness*

We first consider in this study adsorption in a mesoporous system having a nanometric surface roughness. We have considered Xe adsorption at 195 K in the 4.12 nm rough and smooth cylindrical pores shown in Figure 5.1. Figure 5.4 presents GCMC configurations of Xe atoms adsorbed for different pressures in these pores. These simulation snapshots correspond to transversal views (slice) of the adsorbed Xe atoms. We have also reported hydrogen atoms, which delimitate the pore surface. In the case of the rough pore, we have separated:

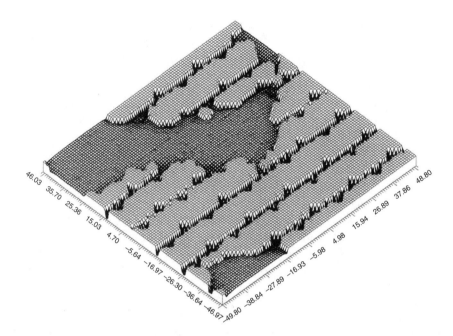

Figure 5.3 A 2D energy grid map. The two in-plane dimensions are x and y space dimensions (in Å) while the third dimension is the adsorbate/matrix potential energy. Darkest areas correspond to lowest adsorbate/matrix potential energy sites. In the mesoporous regions, adsorption sites are located near to the interface in the regions of large curvature. In the microporous regions, the adsorption sites are in the zeolitic channels at well-defined locations.

(*i*) trapped atoms in the troughs of the wall texture, and (*ii*) adsorbed atoms on the remaining surface. An Xe atom located at a z-position corresponding to the *i*th strip is defined "trapped" if its distance to the pore axis is greater than either R_{i+1} or R_{i-1} (the latter are the radius of the $i + 1$th and $i - 1$th strips, respectively). "Adsorbed" atoms are Xe atoms that do not obey to this rule. Xe atoms in the rough/smooth pores do not uniformly cover the pore surface but rather form atomic clusters. Even for high pressures, Xe adsorption does not lead to a flat gas/adsorbate interface. Pellenq et al., have obtained similar simulation results showing that Xe atoms do not wet the Vycor surface but form micro-droplets in the pore regions of highest surface curvature (the other parts of the pore surface being uncovered) [18]. Although Xe atoms adsorb in the primary adsorption sites of lower potential energy at the very first step of the adsorption process, incoming adsorbed atoms tend to aggregate with atoms already adsorbed and, consequently, form clusters (or droplets) located in the regions of space of higher (local) curvature. Interestingly, we have found a different behavior for Ar adsorption at 77 K in a similar rough silica pore: once Ar atoms have filled the troughs, the adsorbate covers uniformly the pore surface [19]. This different "wetting" behavior for Ar and Xe atoms is due to the stronger Xe/Xe interaction than that for Ar atoms. As revealed by snapshots in

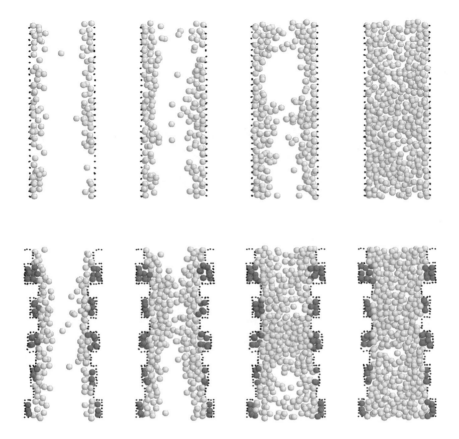

Figure 5.4 (a) Transversal views of Xe atoms adsorbed at 195 K in the 4.12 nm rough pore (bottom) and in the 4.12 smooth pore (top). Gray and white spheres are respectively trapped atoms and adsorbed atoms (*see text*). Black points are hydrogen atoms, which delimitate the pore surface. Pressures are from the left to the right $P = 0.15\ P_0$, $P = 0.26\ P_0$, $P = 0.31\ P_0$, and $P = 0.41\ P_0$.

Figure 5.4, the filling mechanism for both the rough and smooth pore is a continuous process. As the pressure increases, the pore filling proceeds through important density fluctuations along the pore axis, which lead to the presence of gas micro-bubbles enclosed by liquid-like bridges.

Figure 5.5 shows Xe adsorption isotherms at 195 K for the 4.12 nm rough and smooth pores. As expected from simulation snapshots in Figure 5.4, the condensation mechanism in the rough/smooth pore does not correspond to a discontinuous transition between two distinct situations (partially filled and completely filled pores). In addition, we have checked that those mechanisms are reversible. These results show that for this pore size (4.12 nm) the pseudo-critical temperature, defined as the temperature at which adsorption/ desorption hysteresis disappears, is lower than 195 K. The effect of the surface roughness is to shift toward the low-pressure end the filling pressured

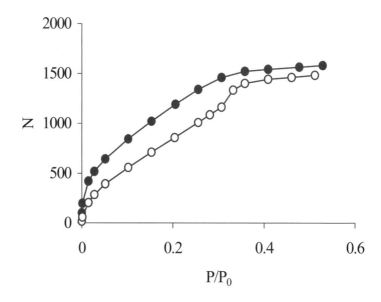

Figure 5.5 Xe adsorption isotherm at 195 K in a 4.12 nm cylindrical pore (empty circles) and in a 4.12 rough pore (filled circles).

of the pore compared to the case of the smooth pore. We have already observed a similar effect in the case of Ar adsorption at 77 K in a pore with morphological defects (pore with a constriction, ellipsoidal pore) [20]. Also, we observe that, due to the surface roughness, the adsorbed amounts in the rough pore are larger than those obtained for the smooth pore. The adsorption branch for the rough pore is smoother than that for the smooth pore. We have already noted [19,21] that the pore size dispersion induced by the morphological disorder (constriction or surface roughness) leads to a dispersion of the filling pressures and, thus, to a smooth adsorption branch. Note that the type of the adsorption isotherm for the rough pore does not correspond to any type in the IUPAC classification [22].

Figure 5.6 shows the adsorbed amounts due to trapped atoms in the troughs of the pore wall. We have also reported the Xe adsorption isotherm obtained at 195 K for the silicalite zeolite sample, which is a purely (ordered) microporous sample. Adsorbed amounts have been normalized to the maximum number of atoms. Both adsorption isotherms correspond to the type I in the IUPAC classification [22], which is usually interpreted as the signature of microporous adsorbents. We observe that for all pressures, the adsorbed amounts for the zeolitic pore are always larger than those obtained for the rough pore. This result is due to the fact that silicalite pores (5 Å) are smaller than the characteristic size of the troughs of the rough pore (10 Å). In Figure 5.7 we show the isosteric heat curve versus the pore filling fraction for the smooth and rough cylindrical nanopores. We have also reported the data for the silicalite sample. The isosteric heat of adsorption for the rough pore at very low coverage (29.0 kJ/mol) is surprisingly close to that obtained

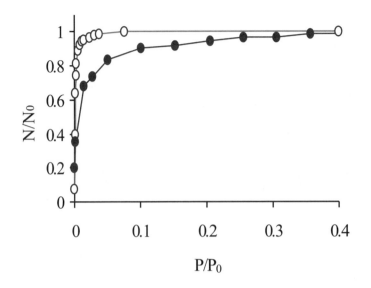

Figure 5.6 Xe adsorption isotherm at 195 K: (filled circles) in the troughs of the 4.12 nm rough pore (i.e., contribution due to trapped atoms). (empty circles) are for Xe adsorption at 195 K in silicalite.

for the silicalite pore (27.0 kJ/mol). However, the overall shape of the isosteric heat of adsorption curves for those two samples is different when considered over the entire adsorption process. In the case of the ordered microporous adsorbent (silicalite), the isosteric heat of adsorption is constant

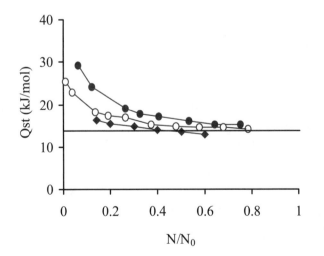

Figure 5.7 Isosteric heat of adsorption versus the pore filling fraction for Xe at 195 K: (empty circles) 4.12 nm smooth pore, (filled circles) 4.12 rough pore, (filled diamonds) experimental data from [12]. The horizontal dashed line indicates the heat of liquefaction of bulk Xe (13.5 kJ/mol).

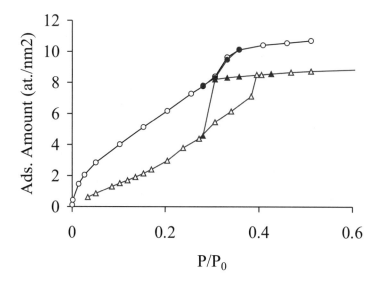

Figure 5.8 Adsorption isotherm of xenon in a purely-mesoporous pseudo-Vycor system at 195 K compared to that in the smooth cylindrical pore.

whatever the filling fraction while it is a decreasing function for the rough pore. This behavior for the isosteric heat in the rough pore is obviously due to the adsorption of Xe atoms in the mesoporous region of the pore.

Figure 5.8 compares Xe adsorption isotherms at 195 K for the purely mesoporous Vycor sample (Figure 5.3) and that for the smooth cylindrical pore (Figure 5.1). For each porous matrix, adsorbed amounts have been normalized to the geometrical surface (Vycor 400 nm^2, smooth cylindrical pore 138 nm^2). The adsorption isotherm for the Vycor sample presents a hysteresis loop corresponding to the irreversibility of the capillary gas-liquid transition in a mesoporous structure (due to metastable states upon both condensation and evaporation). At low pressure, the adsorption isotherm is nearly linear with increasing pressure and can be classified as being of type III in the IUPAC classification [22]. As already mentioned, an analysis of atomic configurations (snapshots) reveals that xenon does not "wet" the inner surface of such mesoporous solids [7,18]. This result is in full agreement with that obtained in the case of xenon adsorption and condensation in the smooth and rough cylindrical mesopores presented here above. At higher pressure, the adsorption/desorption presents a type II hysteresis loop in IUPAC classification as usually found in the experiments for a disordered mesoporous matrix [22]. For the Vycor sample, the average pore size (3.6 nm), given from the first momentum of the chord length distribution, is smaller than the diameter of the smooth cylindrical pore (4.12 nm). It thus may be surprising to find a hysteresis loop for the Vycor sample while the adsorption isotherm for the smooth cylindrical pore is reversible. A possible explanation for this result is the following: capillary condensation occurs in the largest cavity of the Vycor sample, which are domains with a characteristic pore size of 5 nm (as estimated from the chord length distribution calculated by Pellenq and Levitz [7]). Consequently, the hysteresis observed for the Vycor sample is due to the Xe condensation in the largest cavity while the filling of the other regions is reversible. The different adsorption/condensation behaviors for the smooth cylindrical pore and the Vycor structure show that the regular pore model cannot capture the essential features of the adsorption isotherm for complex disordered porous matrix.

It is interesting to see that, once normalized to the surface area, the smooth cylindrical pore adsorbs more Xe atoms that the Vycor-like sample although Vycor exhibits an intrinsic surface roughness. At a first step, one has to inspect for both samples the so-called isosteric heat curve, Qst, that represents the different energetics of the filling process. This quantity has two components: the adsorbate/adsorbate and the adsorbate/substrate contributions. Figure 5.9 presents the isosteric heat curve for the smooth cylindrical pore and the Vycor sample as a function of the gas relative pressure. This type of isosteric heat curve is usually interpreted as being characteristic of adsorption in an energetically heterogeneous environment: the decrease of the isosteric heat as loading increases is due to the decrease of the adsorbate/silica contribution; the adsorbate/adsorbate being of course an increasing function of loading. For both the cylindrical pore and the Vycor sample, the total isosteric heat tends to the enthalpy of liquefaction for Xe at 195 K, Qst = 13.5 kJ/mol. As shown in Figure 5.9, both adsorption in the smooth cylindrical pore or Vycor exhibit the same energetics (within less than one kJ/mol). Two questions now arise: (*i*) how is it possible that a smooth cylindrical surface can adsorb more than the rough one of Vycor and (*ii*) how a smooth

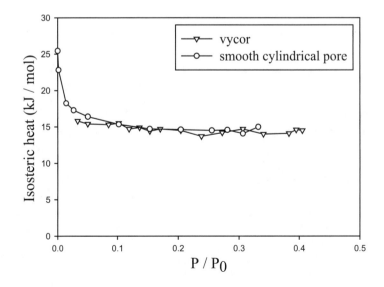

Figure 5.9 Isosteric heat *versus* loading curve at 195 K for Xe/smooth cylindrical pore and the Vycor system, respectively.

cylindrical pore can have a heat curve allure of heterogeneous environment. The question about the roughness of Vycor was addressed many years ago: the intrinsic surface roughness of Vycor was evidenced by small angle neutron scattering experiments showing that the scattered intensity does not obey the Porod law [4,6].

The difference in adsorbed amounts between the smooth cylindrical pore and the Vycor sample can be qualitatively explained by introducing the concept of active surface for adsorption (ASA). In the case of a smooth cylindrical pore, the entire geometrical surface is available and, therefore, active for adsorption. In a disordered material like Vycor, at a first sight, the ASA corresponds to regions of the surface having a positive curvature (concave). As shown in Figure 5.3, the void/matrix interface in Vycor necessarily exhibits more positive curvature than negative curvature. However, the inner Vycor surface also exhibits regions with negative curvature (convex) that explain the smaller affinity with the fluid for this disordered porous structure compared to the smooth cylindrical pore. Consequently, this may explain why the number of atoms adsorbed per unit of surface (using the geometrical surface area) is smaller for the Vycor sample. The different adsorbed amounts for the Vycor sample and the cylindeical pore (Figure 5.8) are thus due to different surface curvatures of the inner surface and not to the surface chemistry that is identical for both samples. This is confirmed by the fact that the heat of adsorption as a function of the chemical potential is the same for the Vycor sample and the smooth cylindrical (Figure 5.9). One may define the ASA as the part of the geometrical surface that is actually involved in the adsorption of the first atoms. As mentioned above, the ASA for the Vycor sample is *a priori* made of the surface regions having a positive curvature and, thus, should have a similar adsorption capacity (atoms per unit of surface) to that for the smooth cylindrical pore. On the basis of this definition, we have estimated that the ASA for the Vycor sample is about 81 m^2/g, i.e., 37%, of the geometrical surface. It is striking that this value is in a very good agreement with the BET surface area measured from the simulated Xe adsorption isotherm (86 m^2/g) [7] also in full agreement with the experimental value determined from xenon adsorption [12] at 195 K. A further analysis of such an adsorption process in Vycor has shown that the specific surface area as measured from xenon adsorption isotherm at 195 K was underestimated by a factor of two compared to the geometrical value obtained from the chord length distribution [7,18] in agreement with experiment on real disordered silica mesoporous solids [12]. Note that the ASA is an adsorbate dependent concept, i.e., its value depends on the "wetting" behavior of the adsorbed atoms. For instance, we expect the ASA for Ar atoms to be larger than that for Xe atoms since the Ar atoms are less sensitive to the negative curvature of the Vycor surface and tend to uniformly cover the pore wall (see discussion above). Assuming that the BET method provides an estimation of the ASA, the different wetting behavior for Xe and Ar atoms may explain the much higher BET surface assessed from Ar adsorption (145 m^2/g) compared to that obtained from Xe adsorption (86 m^2/g).

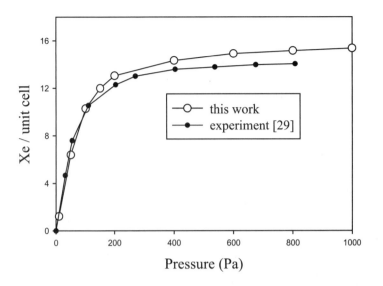

Figure 5.10 Simulated adsorption isotherm of xenon in silicalite at 195 K compared to experiment.

5.3.2 Xe adsorption and condensation in a vycor-like having a mixed micro and mesoporosity

We now consider a mesoporous Vycor having a microporous texture (zeolitic channels) as described earlier. All results presented in this section have been obtained using the "true" two-body potential for the Xe-Xe interaction. Figure 5.10 shows the GCMC adsorption isotherm of Xe in silicalite at 195 K, which is in good agreement with experimental data [16]. The adsorption isotherm is reversible and characteristics of microporous solids (type I in the IUPAC classification [22]). Note that in the case of adsorption in silicalite, there is no adsorbate/hydrogen interaction to consider since silicalite is a pure silica structure and the adsorbate/matrix potential used throughout this work is the same as far as oxygen and silicon species are concerned. A statistical sampling of the porosity of silicalite obtained by probing the adsorbate/silicalite potential energy grid, gives a porosity in the case of Xe at around $\phi_{silicalite} = 12.4\%$. The maximum adsorbed amount is around 16 Xe per unit cell: this corresponds to a density of 0.0240 Xe/A^3. The xenon adsorbed phase in silicalite is much denser that in the liquid bulk phase at the same temperature (0.0129 Xe/A^3). This effect was also observed in the case of argon confined in silicalite at 77 K [23]: 0.0467 Ar/A^3 (in silicalite confined), 0.0232 Ar/A^3 (bulk solid). This shows that the Gurvitch rule [22], which states that the confined fluid has the same density as the bulk liquid, is not valid. It is worth noting that recent molecular simulations of nitrogen adsorption at 77 K in realistic microporous carbons have reached the same conclusion concerning the Gurtvich rule [24].

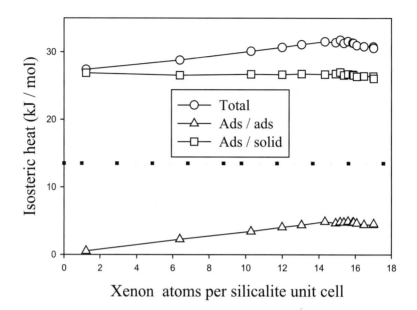

Figure 5.11 Isosteric heat versus loading curve for the Xe/silicalite system at 195 K. Its two contributions (ads/ads/ and ads/solid) are also shown. The horizontal dashed line indicates the value of the heat of liquefaction of bulk xenon (13.5 kJ/mol).

Figure 5.11 presents for the silicalite sample the isosteric heat of adsorption versus the filling fraction of the porosity. This curve is characteristic of adsorption on an energetically homogeneous hypersurface of potential energy: the adsorbate/zeolite contribution remains constant as loading increases. The total isosteric heat being the sum of the adsorbate/adsorbate and the adsorbate/surface terms, is then an increasing function of loading since the adsorbate/adsorbate contribution also increases with loading. However, note that at loading corresponding to 16 Xe per unit cell, the total isosteric heat curve presents a slight decrease, which allows one to locate the maximum amount that can be adsorbed in silicalite zeolite. These features on the isosteric heat curve have been also reported for Xe and CH_4 adsorption in NaY zeolite [14].

We now consider adsorption in the pure Vycor mesoporous structure as obtained from the off-lattice method (see figure 5.2). Figure 5.12 presents the Xe adsorption isotherm at 195 K. This curve has been obtained with the "true" Xe-Xe pair potential as in the case of the silicalite sample and, thus, differs from the Xe adsorption isotherm for the same Vycor sample which has been obtained using the effective Xe/Xe pair potential (shown in figure 5.8).

We have given the adsorption characteristics curves (adsorption isotherm and isosteric heat curve) for both silicalite zeolite and Vycor. We now consider a porous matrix having both microporous and mesoporous region. As explained earlier the mixed porous system was obtained by applying the off-lattice method for the reconstruction of Vycor to a simulation box originally containing 5*5*7 unit cells of orthorhombic silicalite [9]. Figure 5.13

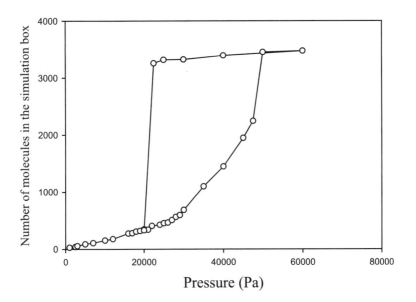

Figure 5.12 Adsorption of xenon in the pure Vycor system at 195 K (note the Xe-Xe potential is that used in the case of silicalite).

presents the Xe adsorption/desorption isotherm at 195 K in such a mixed porous material. One can see that this curve still exhibits the hysteresis phenomenon characteristic of capillary condensation (this point will be discussed below). One can see that in the low-pressure range, the shape of the Xe adsorption isotherm for the mixed porous material closely resembles that for the pure zeolitic material (see Figure 5.10). This is obviously due the presence of microporous regions embodied in the core of this essentially mesoporous material. The density of the matrix was evaluated at 1.249 g/cm³ which corresponds to a mesoporosity at $\phi_{meso} = 30\%$ ($\rho_{mixed} = 0.3* \rho_{silicalite}$; $\rho_{silicalite} = 1.785$ g/cm³ [9]). However, if the density of the mixed material is compared to that of non-porous silica $\rho_{silicalite} = 2.15$ g/cm³, the porosity of the mixed material is found to be 42%.

Figure 5.14 presents the corresponding isosteric heat curve. Two regions can be distinguished. The first region corresponds to adsorption within the microporosity. Indeed, the shape of the isosteric heat curve is the same as that obtained for the pure silicalite sample (see Figure 5.11). In particular, the maximum adsorbed amount in the microporosity can be identified by locating the abrupt decrease at around 1900 Xe. This corresponds to a density of 0.0234 Xe/A³ (very close to that found in the case of xenon adsorption in pure silicalite at the same temperature). This density value is obtained by multiplying the total simulation box volume (938917 A³) by the factor [$\phi_{silicalite}* (1-\phi_{meso})$] in order to have the microporous volume of our numerical mixed porosity sample (81207 A³). Therefore, we infer that the analysis of the isosteric heat curve of a mixed porous material is an efficient tool to calculate a microporous volume rather than analysing the adsorption

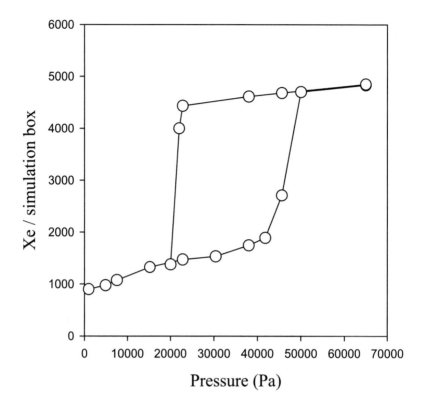

Figure 5.13 Adsorption of xenon in the mixed-porosity system at 195 K (note the Xe-Xe potential is that used in the case of silicalite).

isotherm, which has the almost same shape in the case of a microporous environment or a rough surface. This calorimetry-based method is especially valid when the energetics contrast between adsorption in the micro and mesoporous regions is large as is the case for xenon adsorption in silica pores. Note that the isosteric heat of adsorption after capillary condensation remains to a value being about 4 kJ/mol higher than the enthalpy of lique-faction by contrast to that obtained in the case of the smooth and rough cylinders (see above). This is clearly the consequence of the different adsorbate-adsorbate potential functions used in the two studies. We have used in the case of the mixed silicalite/Vycor system the potential function which describes the best the fluid confined in the zeolitic region; this potential that we called "true two-body" potential being more appropriate to describe xenon in low coordination environment.

The low-pressure snapshot presented in Figure 5.15a does indeed confirm the overall xenon adsorption mechanism: the vast majority of the xenon atoms are adsorbed in the microporous channels of silicalite; there are very few xenon atoms adsorbed in the vicinity of the micropore openings (connections to mesoporous domains). In the case of argon adsorption, the isosteric heat of adsorption at 77 K is 14,5 kJ/mol when adsorbed in silicalite [11,13]

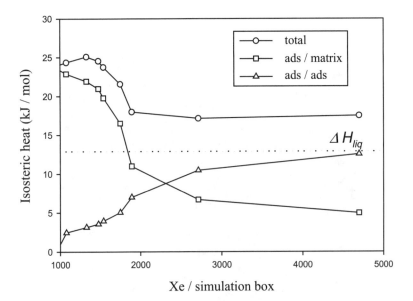

Figure 5.14 Isosteric heat versus loading curve for the Xe/mixed-porosity system at 195 K. Its two contributions (ads/ads/ and ads/solid) are also shown. The horizontal dashed line indicates the value of the heat of liquefaction of bulk xenon (13.5 kJ/mol).

while it is around 13 kJ/mol in mesoporous silica [7]. Therefore one expects that adsorption in the microporosity and in the mesoporosity will occur on the same pressure range. As a consequence, the isosteric heat curve will not distinguish between these two processes. Interestingly, the adsorbed amount required to fill the microporous part is reached just at the onset of the capillary condensation in the mesoporosity. This result confirms that there is no Xe adsorbed film inside the mesoporosity; Xe gradually condenses in the high curvature regions of the mesoporous interface (see Figure 5.15b and 5.15c).

The ratio between the microporous volume of the mixed material with that of pure silicalite ($\phi_{silicalite} V_{silicalite}$) is 78.7. The original Xe/silicalite adsorption isotherm multiplied by this number 78.7 is thus the microporous contribution to the total adsorption isotherm for the mixed porous material. The addition of such a contribution to the pure mesoporous isotherm gives a composed (micro/meso) isotherm shown in Figure 5.16. One can see that the composed adsorption isotherm is in good agreement with that directly obtained for the mixed porous material. This further demonstrates that adsorption and condensation for Xe atoms proceed in two distinct steps: (*i*) filling of the microporosity and (*ii*) adsorption/condensation in the mesoporosity. Thus, the adsorption of Xe at 195 K seems to provide an interesting method that enables to distinguish micro and mesoporosity for silica nanopores. We further stress that it may not be applicable to other simple usual gases such as argon or nitrogen at 77 K since the energetics contrast between the filling of the micro- and meso-porous regions may not be sufficient.

(a)

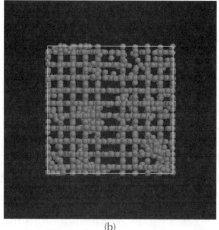

(b)

Figure 5.15 (a) Xe adsorbed phase (x,y) snapshot at low pressure, (b) *idem* just before capillary condensation (c) *idem* at mesopore completion.

We now discuss the hysteresis phenomenon observed for both the pure mesoporous and the mixed porous materials. Capillary condensation is often seen as a first order transition: theoretical and simulation studies have demonstrated that it is indeed the case for simple pore geometries where no disorder is present (neither in terms of pore morphology nor in terms of network topology). However, the possibility of having no first order phase transition for disordered systems is now considered in some cases (even for

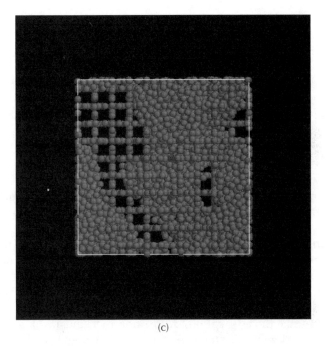

(c)

Figure 5.15   *(Continued).*

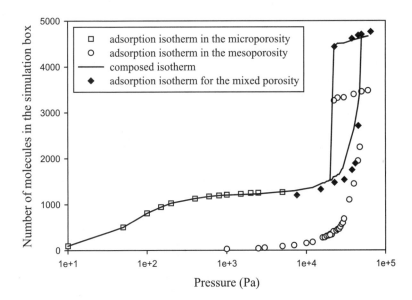

Figure 5.16   Comparison between the composed isotherm and that obtained directly from GCMC calculations of the mixed porosity material.

adsorption/desorption isotherms presenting the hysteresis phenomenon): due to the collection of different curvatures, the system possesses a large number of metastable states of roughly the same grand free energy. The existence of such a complex energy landscape tends to avoid phase coexistence [25]. However, the hysteresis loop observed in experiments for disordered systems does have the usual behaviour found in the case of simple pore geometries (such as slit pores) for which capillary condensation/evaporation is a true first order transition: it shrinks and disappears at a temperature that we defined as a pseudo or apparent critical temperature [26]. If one postulates that there is indeed a pore critical point, then at constant temperature, a confined fluid will be supercritical in small pores and there will be a pore critical size above which the fluid remains sub-critical. In terms of confinement, we can thus consider a fluid confined in a zeolitic microporous network to be supercritical (having large density fluctuations). If a given mesopore is connected to another one through some micropores, diffusion between mesopores can always be achieved thanks to the micropores since there will be no gas-liquid interface. In other words, a given mesopore can always empty through smaller pores filled with supercritical fluid. However, the critical property of such a confined fluid may not show up since the adsorption/desorption mechanisms in a nanometric confinement is triggered by the solid/fluid interaction. The thermodynamics, dynamics and structure of the adsorbed phase are dominated by the solid/fluid contribution to the total energy (even in the high loading regime). As shown in Figure 5.13, the presence of a hysteresis loop in the adsorption/desorption isotherm seems to indicate that mesoporous part of the mixed material is still below the pseudo-critical point. The microporosity seems to have no influence on the capillary phenomenon that occurs in the mesoporosity.

Monson and Sarkisov [27] have shown that the GCMC approach to simulate adsorption and desorption processes is strictly equivalent to brutal force (dual controlled volume) molecular dynamics (GCMD): adsorption/desorption isotherms for a Lennard-Jones fluid confined in a disorder mesoporous material are identical with both methods. Thus, the sampling of the phase space in the GCMC scheme is not at fault and does nicely describe capillary condensation. However, one may legitimately think that some cautions must be taken because the standard GCMC algorithm allows the creation and destruction of particles anywhere in the system "bypassing" pore constrictions. Indeed, GCMD results for a Lennard-Jones fluid, confined in an ideal model describing a mixed porous material made of small micropores connected to a larger pore, have shown that pore-blocking effects can occur upon melting of the adsorbed phase in the microporosity leading to pore blocking effects [28]. In the case of real mesoporous materials having an additional microporous texture, such an effect is not relevant since access to the mesoporous region does not depend on diffusion through the microporosity. In fact, the converse is more probable: mesoporosity makes easier the access to microporosity.

5.4 Conclusion

We have performed Grand Canonical Monte Carlo simulations of Xe adsorption at 195 K in silica mesoporous matrices having either a microporous structure or a nanometric surface roughness. The purpose of this study is to gain some insights on the influence of the surface texture on the adsorption/condensation phenomena. We have first considered the adsorption in a rough cylindrical pore composed of strips of different diameters (the mean pore size is 4.12 nm and the size dispersion about 1nm). The results for this rough pore have been compared with those obtained for a smooth pore having a diameter of 4.12 nm. Both Xe adsorption isotherms are reversible for the temperature at which our simulations were performed, 195 K. The filling pressure of Xe atoms in the rough pore is lower than that obtained for the smooth cylindrical pore. Also, we have found that the morphological disorder of the rough pore (i.e., varying pore diameter) leads to an adsorption branch much smoother than that obtained for the regular cylindrical pore.

We have also compared the results for the smooth cylindrical pore and those obtained for a realistic model of porous glass Vycor. We show that the Xe adsorption isotherm at 195 K for a simple model, i.e., a regular cylindrical pore, cannot reproduce the main features of the adsorption isotherm for the disordered porous matrix Vycor. In particular, the capillary condensation for the Vycor sample (mean pore size 3.6 nm) is found to be irreversible while the adsorption isotherm for the 4.12 nm cylindrical pore exhibits no hysteresis loop (the temperature is below the pseudocritical point for this pore size). The metastability-driven irreversibility observed for the filling/emptying of the Vycor porous sample is explained as follows: the capillary condensation/evaporation occurs in the largest cavity of the disordered structure having a size larger than 4.12 nm for which the pseudo-critical temperature is above the temperature of our simulations (195 K). The adsorbed amounts per unit of geometrical surface are found to be larger for the smooth cylindrical pore than those for the Vycor sample. This result is explained by introducing the concept of *Active Surface for Adsorption* (ASA). The ASA represents the part of the geometrical surface that is actually involved in the adsorption of the first adsorbate atoms. In the case of the cylindrical pore the ASA equals the geometrical pore surface that has only a positive curvature (preferential adsorption sites). In contrast, the Vycor inner surface possesses both regions of positive and negative curvature. Consequently, the ASA for the disordered porous structure necessarily is lower than the geometrical surface. Note the value of the ASA is adsorbate dependent and is expected to strongly depart from the geometrical surface for adsorbate, such as Xe, that does not uniformly cover the inner surface. Interestingly, we have found that the ASA surface is very close to the surface assessed using the BET method (in which it is assumed that the adsorbate forms a uniform monolayer on the pore wall).

In a second set of simulations, we have simulated the Xe adsorption isotherm at 195 K for a disordered porous material having both mesoporous and microporous regions. The latter has been obtained by convoluting the functional (giving the distribution of the porous voids) of the purely meso-porous Vycor sample and that for a zeolite silicalite microporous adsorbent. The Xe adsorption/desorption isotherm at 195 K for this mixed micro/meso-porous material shows that the difference of energetics between zeolitic micropores and CPG mesopores leads to two distinct adsorption processes which occur consecutively: (*i*) micropore filling and (*ii*) adsorption/conden-sation in the mesoporous regions. As a consequence, both the microporous and the mesoporous can be assessed independently. In particular, we show that the adsorption isotherm for the mixed structure is equivalent to the sum of the adsorption isotherm for a pure silicalite sample and that for the pure mesoporous Vycor sample. We suggest that the use of xenon at 195 K may be an efficient way to distinguish between microporous and mesoporous volumes, which can be directly evaluated thanks to a straightforward anal-ysis of the isosteric versus loading curve.

Acknowledgments

The work is supported by the Institut du Développement et des Ressources en Informatique Scientifique, (CNRS, Orsay, France): T3E computing grants n° 991153 and n° 0211427.

References

1. L. D. Gelb, K. E. Gubbins, R. Radhakrishnan, M. Śliwinska-Bartkowiak, "Phase separation in confined geometries," *Rep. Prog. Phys.* 62, 1573–1659 (1999).
2. A. Berenguer-Murcia, J. Garcia-Martinez, D. Cazorla-Amoros, A. Martinez-Alonso, J. M. D. Tascon and A. Linares-Solano, "About the exclusive meso-porous character of MCM-41," in *Studies in Surface Science and Catalysis*, Vol. 144, (F. Rodriguez-Reinoso, F.; McEnaney, B.; Rouquerol, J.; Unger, K. K.; Eds.) Elsevier Science, 83–90 (2002).
3. S. H. Joo, R. Ryoo, M. Kruk, and M. Jaroniec, "Evidence for general nature of pore interconnectivity in 2-dimensional hexagonal mesoporous silicas pre-pared using block copolymer templates," *J. Phys. Chem.* B, 106, 4640–4646 (2002). S. Jun, S. H. Joo, R. Ryoo, M. Kruk, M. Jaroniec, Z. Liu, T. Ohsuna and O. Terasaki, "Synthesis of New, Nanoporous Carbon with Hexagonally Ordered Mesostructure," *J. Am. Chem. Soc.* 122, 10712–10713 (2000).
4. P. E. Levitz, G. Ehret, S. K. Sinha, and J. M. Drake, "Porous Vycor glass: the microstructure as probed by electron microscopy, direct energy transfer, small-angle scattering and molecular adsorption," *J. Chem. Phys.* 95, 6151–6161 (1991).
5. D. H. Olson, G. T. Kokotailo, S. L. Lawton, and W. M. Meier, "Crystal structure and structure-related properties of ZSM-5," *J. Phys. Chem.* 85, 2238–2243 (1981).
6. P. E. Levitz, "Off-lattice reconstruction of porous media: critical evaluation, geometrical confinement and molecular transport," *Adv. Coll. Int. Sci.* 76–77, 71–106 (1998).
7. R. J. M. Pellenq, and P. E. Levitz, "Capillary condensation in a disordered mesoporous medium: a grand canonical Monte Carlo study," *Mol. Phys.* 100, 2059–2077 (2002).

8. M. J. Torralvo, Y. Grillet, P. L. Llewellyn, and F. Rouquerol, "Microcalorimetric study of argon, nitrogen, and carbon monoxide adsorption on mesoporous Vycor glass," *J. Coll. Int. Sci.* 206, 527–532 (1998).

9. E. M. Flanigen, J. M. Barret, R. W. Grose, J. P. Cohen, R. L. Patton, R. M. Kirschner, and J. V. Smith, "Silicalite, a pure siliceous form of ZSM-5 zeolite," *Nature* 271, 512–514 (1978).

10. P. L. Llewellyn, J. P. Coulomb, Y. Grillet, J. Patarin, G. Andre, and J. Rouquerol, "Adsorption by mfi-type zeolites examined by isothermal microcalorimetry and neutron diffraction. ii: nitrogen and carbon monoxide," *Langmuir* 9, 1852–1856 (1993).

11. R. J.-M. Pellenq, and D. Nicholson, "Intermolecular potential function for the physical adsorption of rare gases in silicalite," *J. Phys. Chem.* 98, 13339–13349 (1994).

12. C. G. V. Burgess, D. H. Everett, and S. Nuttal, "Adsorption of CO2 and xenon by porous glass over a wide range of temperature and pressure: applicability of the Langmuir case VI equation," *Langmuir* 6, 1734–1738 (1990).

13. R. J. M. Pellenq, and D. Nicholson, "Grand canonical Monte-Carlo simulation of adsorption of small molecules in silicalite zeolite," *Langmuir* 11, 1626–1635 (1995).

14. R. J.-M. Pellenq, B. Tavitian, D. Espinat, and A. Fuchs, "Grand canonical Monte-Carlo simulations of adsorption of polar and non polar molecules in NaY zeolite," *Langmuir* 12, 4768–4783 (1996).

15. J. A. Barker, R. O. Watts, J. K. Lee, T. P. Schafer, and Y. T. Lee, "Interatomic potentials for krypton and xenon," *J. Chem. Phys.* 61, 3081–3089 (1974).

16. D. A. Kofke, "Semigrand canonical Monte Carlo simulation; Integration along coexistence lines," *Adv. Chem. Phys.* 105, 405–441 (1999).

17. D. Nicholson and N. G. Parsonage in *Computer Simulation and the Statistical Mechanics of Adsorption,* Academic Press (1982).

18. R. J.-M. Pellenq, S. Rodts, V. Pasquier, A. Delville, and P. E. Levitz, "A Grand Canonical Monte-Carlo simulation study of xenon adsorption in a Vycor," *Adsorption* 6, 241–249 (2000).

19. B. Coasne, A Grosman, C. Ortega, and R. J.-M. Pellenq, "Physisorption in nanopores of various sizes and shapes: A grand canonical Monte Carlo study," in *Studies in Surface Science and Catalysis,* Vol. 144, (F. Rodriguez-Reinoso, F.; McEnaney, B.; Rouquerol, J.; Unger, K. K.; Eds.) Elsevier Science, 35–42 (2002).

20. B. Coasne and R. J.-M. Pellenq, "Grand canonical Monte Carlo simulation of argon adsorption at the surface of silica nanopores: Effect of pore size, pore morphology and surface roughness," *J. Chem. Phys.,* in press (2004).

21. B. Coasne and R. J.-M. Pellenq, "A grand canonical Monte Carlo study of capillary condensation in mesoporous media: from a single regular pore to a disordered porous matrix," *J. Chem. Phys.,* submitted (2003).

22. F. Rouquerol, J. Rouquerol and K. Sing, in *"Adsorption by Powders and Porous Solids,"* Academic Press (1998).

23. D. Douguet, R. J.-M. Pellenq, A. Boutin, A. H. Fuchs, and D. Nicholson, "The adsorption of argon and nitrogen in silicalite-1: A Grand Canonical Monte-Carlo study," *Mol. Sim.* 17, 255–267 (1996).

24. J. Pikunic, C. Clinard, N. Cohaut, K. E. Gubbins, J. M. Guet, R. J. M. Pellenq, I. Rannou, and J. N. Rouzaud, "Structural modeling of porous carbons: constrained reverse Monte Carlo method," *Langmuir* 19, 8565–8582 (2003).

25. E. Kierlick, M.-L. Rosinberg, G. Tarjus, and P. Viot, "Equilibrium and out-of-equilibrium (hysteretic) behavior of fluids in disordered porous materials: Theoretical predictions," *Phys. Chem. Chem. Phys.,* 3, 1201–1206 (2000).

26. R. J.-M. Pellenq, B. Rousseau, and P. E. Levitz, "A Grand Canonical Monte-Carlo Study of argon adsorption/condensation in mesoporous silica glasses," *Phys. Chem. Phys.* 3, 1207–1212 (2001).
27. L. Sarkisov, and P. A. Monson, "Hysteresis in Monte-Carlo and molecular dynamics simulations of adsorption in porous materials," *Langmuir* 16, 9857–9860 (2000).
28. M. W. Maddox, K. E. Gubbins, and N. Quirke, "A molecular simulation study of pore networking effects," Mol. Sim. 19, 267–272 (1997).
29. D. Shen, Ph. D. Thesis, Imperial College, University of London, 1992.

chapter six

Molecular simulation of adsorption of guest molecules in zeolitic materials: a comparative study of intermolecular potentials

A. Boutin*

S. Buttefey

A. H. Fuchs

A. K. Cheetham

University of California

Contents

*Corresponding author.
Reprint from *Molecular Simulation*, 27: 5–6, 2001. http://www.tandf.co.uk

6.1   Introduction

Zeolitic materials and related open-framework inorganic materials are gaining increasing importance in industrial applications. In two of the most widespread applications, i.e., molecular sieving and catalysis, a crucial role is played by adsorption and transport of the guest molecules. From a more academic point of view, the behavior of fluids in confined geometries has also attracted much interest in the past few years. While the macroscopic science of this field is well developed, there is a need for a more fundamental microscopic understanding of the phenomena, as well as means for predicting thermodynamics and transport properties in a variety of guest-host systems.

Molecular simulation, in conjunction with experiments, has played an important role in the past few years in developing our understanding of the relationship between microscopic and macroscopic properties of confined molecular fluids in zeolitic materials.

Two principal types of theoretical treatments of guest molecules in zeolite hosts can be found in the literature. On one hand, *ab initio* quantum chemistry techniques are used to address the problem of molecular chemisorption processes and reactions at Brønsted acid sites. On the other hand, classical Monte Carlo (MC)/Molecular Dynamics (MD) simulations are used to study adsorption and transport of molecules in zeolite pores.

The quantum chemistry approach is rather time consuming and, for this reason, calculations were often limited in the past to finite cluster models of zeolite. Modern *ab initio* MD codes can now be used to study larger systems, such as a methanol molecule interacting with the Brønsted site in a periodic model of chabazite [1,2].

The classical MC/MD approach has been widely used to study the behaviour of simple molecules (e.g., rare gases or simple hydrocarbon molecules) in silicious zeolites such as silicalite. A large number of different equilibrium configurations of the system can be generated through these techniques, enabling one to compute ensemble average quantities that can be related to thermodynamics and transport properties of the guest molecules. This approach relies on semi-empirical intermolecular potentials, which constitutes a main drawback of the classical methods.

Bridging the gap between the quantum chemistry and the classical approaches is a major challenge in molecular simulation. Electronic density functional theory based MD codes are still far from being able to address long time diffusion and high loading adsorption processes. This is not only a problem of computing time. Basic problems remain open, such as the long range dispersion interaction between species which are still not properly handled in the theory. The development of mixed quantum/classical methods, once the embedding problems are solved, is expected to yield new powerful methods for zeolite catalysis studies.

For the time being, the classical, semi-empirical, approach is the only feasible way of addressing thermodynamic and transport phenomena in complex guest/host systems in which no chemical reactivity takes place.

Recently developed techniques allow the simulation of systems that a few years ago were considered impossible to study via computer simulation. Systems of relevance to commercial applications, such as normal and branched alkanes [3], benzene [4], alkyl benzene isomers [5] and halocarbon molecules [6] in aluminosilicate hosts are now being studied by molecular simulation. The MC algorithms used to obtain reliable thermodynamic data (adsorption isotherms and heats) as well as the newly developed semi-empirical forcefields which allow a better transferability of the parameters from one guest/host system to another have been reviewed recently [7].

In this chapter, we address the specific question of reliability and transferability of the forcefield. Most authors agree on the fact that there is no single optimum forcefield for predicting adsorption properties. Some of them believe in an "engineering" approach in which the simplest (and cheapest) potential function should be used, and the potential parameters be readjusted whenever a quantitative prediction is needed. Others try to develop new strategies to derive semi-empirical potentials on a firmer basis. The performance of these two approaches is compared here, for several guest/host systems: argon and methane/$AlPO_4$-5, xylene isomers/faujasite. It is shown that simple, Kiselev type, forcefields can do a good job for the so-called "simple" systems (small guest molecules in a neutral framework). Full scale potentials are needed, however, to model complex systems. These latter forcefields allow a better transferability of the parameters from one system to another, and still make use of a limited number of adjustable parameters.

6.2 Computational methodologies

6.2.1 Monte Carlo (MC) simulations

MC simulations are particularly convenient for computing equilibrium thermodynamic quantities such as the average number of adsorbed molecules $\langle N \rangle$, the isosteric heat of adsorption q_{st}, and the Henry's constants K. In addition, MC simulations provide detailed structural informations, in particular the location and distribution of adsorbed molecules in the pores.

Adsorption quantities have been computed in the Grand Canonical (GC) statistical ensemble in which the chemical potential (μ), volume (V) and temperature (T) are fixed [8]. These thermodynamic conditions are close to the experimental conditions where one wants to obtain information on the average number of particles in the porous material as a function of the external conditions. At equilibrium, the chemical potentials of the fluid bulk phase and the adsorbed phase are equal. The pressure in the reservoir fluid can be calculated from an equation of state, and it is thus directly related to the chemical potential in the adsorbed phase. The ensemble average number of molecules in the zeolite, $\langle N \rangle$, is computed directly from the simulation. By performing simulations at various chemical potentials, at a given temperature, one obtains the adsorption isotherm. Experimental adsorption isotherms yields the *excess* number of molecules adsorbed in the porous medium which is not, in principle, directly comparable to $\langle N \rangle$. Since zeolite pores are small, the correction is negligible under normal conditions.

A full development of the statistical mechanics of the μVT ensemble, the description of the corresponding Monte Carlo algorithms, and the bias MC moves that have been suggested in order to increase the acceptance probability of the insertion/deletion step for anisotropic molecules such as xylene isomers, have been given in several publications, e.g., [8–10] and will not be repeated here.

6.2.2 Potential energy models

The potential energy model is an important input to a molecular simulation. Although the guest–host interaction is the most significant part of the total potential energy, some attention should be paid to the accuracy of the guest–guest interaction as we will see in the discussion section.

The zeolite is usually modeled as a rigid crystal. Following Kiselev [11], most authors have used a rather simplified guest–host potential function U_{GH}. In this model, a Lennard–Jones repulsion–dispersion term acts between the atoms of the guest adsorbate (G) and the oxygen atoms (O) and M$^+$ cations of the host material (H). In the case where the adsorbate molecules are multipolar, an electrostatic term is added, which acts between all atoms of the zeolite (Oxygen, tetrahedrally coordinated (T) atoms and M$^+$ cations) and of the adsorbate:

$$U_{GH}^{Kiselev} = \sum_{G,H \in O,M^+} 4\varepsilon^{GH} \left\{ \left(\frac{\sigma^{GH}}{r_{GH}} \right)^{12} - \left(\frac{\sigma^{GH}}{r_{GH}} \right)^6 \right\} + \sum_{G,H \in T,O,M^+} \frac{q^G q^H}{r_{GH}} \quad (6.1)$$

where r_{GH} is the distance between the guest G and the host H atoms, and q^G and q^H are the partial charges borne by the guest and host atoms, respectively. Some authors have used formal charges for the framework atoms, but usually partial charges are assigned using some quantum chemistry calculation. In most studies, the other part of the potential function, i.e., the host–host interaction, takes the form of an effective two-body potential derived from bulk simulations. In both the guest–host and guest–guest potentials, the well-known Lorenz–Berthelot combining rules are used to handle cross interactions.

Pellenq and Nicholson have developed a full scale guest–host semi-empirical potential (hereafter called the PN potential) with the aim of obtaining a more accurate and transferable potential model [12–20]

$$U_{GH}^{PN} = U_{el} + U_{pol}^{PN} + U_{disp}^{PN} + U_{rep}^{PN} \quad (6.2)$$

in which the first term is the electrostatic interaction calculated in the same way as in the Kiselev potential. Induced interactions, due to partial charges of the framework species, are calculated using the first term of the multipole expansion

$$U_{pol}^{PN} = -\frac{1}{2}\sum_{i \in G} \alpha_i E_{T,O,M^+}^2 \qquad (6.3)$$

where α_i, is the dipole polarizability of atom i of a guest molecule and E is the electrostatic field at the position occupied by atom i due to the partial charges carried by all the host species. The back-polarization and higher order terms are neglected.

The dispersion interaction includes the r^{-6}, r^{-8} and r^{-10} terms, as well as the three-body dispersion Axilrod–Teller term:

$$U_{disp}^{PN} = \left[\sum_{G,H \in T,O,M^+} -\frac{C_6^{GH}}{r_{GH}^6} - \frac{C_8^{GH}}{r_{GH}^8} - \frac{C_{10}^{GH}}{r_{GH}^{10}} \right] + U_{3\,body} \qquad (6.4)$$

where the three-body interaction involving triplets of species i, j and k can be expressed in terms of geometrical factors W^{ijk} and electronic functions Z^{ijk} in the following general form:

$$U^{ijk}(l_1,l_2,l_3) = \sum_{l_1}\sum_{l_2}\sum_{l_3} Z^{ijk}(l_1,l_2,l_3)W^{ijk}(l_1,l_2,l_3) \qquad (6.5)$$

The dispersion coefficients in Equation 6.4 are estimated from a knowledge of the dipole polarizabilities and the partial charges of all interacting species. The calculation of the three-body dispersion term requires the same set of parameters as the two-body terms. All atoms of the framework are considered here, not only the oxygen atoms as in the Kiselev potential.

Finally, the repulsion interaction is represented with an exponential Born-Mayer term:

$$U_{rep}^{PN} = \sum_{G,H \in T,O,M^+} A^{GH}\exp(-b^{GH}r_{GH}) \qquad (6.6)$$

In this latter case, Böhm–Ahlrichs combining rules are used to handle cross interactions:

$$A^{ij} = (A^{ii}A^{jj})^{1/2}, \quad b^{ij} = \frac{2b^{ii}b^{jj}}{b^{ii} + b^{jj}} \qquad (6.7)$$

Usually, the repulsive interaction between the adsorbate and the T atoms can be neglected since the guest molecules are only sensitive to the repulsion from the oxygen atom and the extra-framework cations (when they are present).

Although the PN potential is more sophisticated and rather more time consuming to use in simulations than the Kiselev potential, it should be stressed here that no extra adjustable parameters are introduced in the full scale PN potential. Since the dispersion coefficients C_{6-10}^{GH} are calculated from a knowledge of dipole polarizabilities and the effective number of electrons of the interacting species, the only adjustable parameters are the A's and b's of the repulsion energy.

The PN potential function has been successfully used in a variety of systems: rare gases and small molecules in silicalite [12–16], faujasite [21] and $AlPO_4$-5 [22,23], and benzene and xylene molecules in faujasites [5,10,21]. In the following section, we revisit some of the key results obtained by using different types of potential models. We then discuss the usefulness of either using the simple effective model or the full scale model. We also suggest some further developments in the methods used for modeling adsorption of complex mixtures (such as water+hydrocarbons) in cationic zeolites, which constitutes a major challenge for the future.

6.3   Results

6.3.1   Adsorption of argon in $AlPO_4$-5

The aluminophosphate $AlPO_4$-5 is a microporous crystal with a neutral framework. It consists of alternate tetrahedral aluminium and phosphorus atoms bridged by oxygen atoms. The crystalline lattice is hexagonal space group *P6cc* for ordered Al and P, and was taken from X-ray scattering and neutron powder diffraction studies. The full model has been described in earlier publications [22,23]. The micropores are not interconnected and form unidimensional channels of ~7.3 Å diameter parallel to the crystallographic *c* direction (AFI structural network). The rather simple $AlPO_4$-5 inner surface consists of a regular hexagonal array of oxygen atoms.

The computed adsorption isotherms of argon in $AlPO_4$-5 are shown in Figure 6.1, where they are compared to experiments. The first set of data (*K* for Kiselev-like potential) was obtained using oxygen–argon (O–Ar) Lennard–Jones parameters (see Table 6.1, parameters set 1) obtained from a combination of Smit et al.'s oxygen–alkane potential [24,25] and argon–argon potential parameters [26] using Lorentz–Berthelot combination rules. A rather good agreement is obtained between simulation and experiments. Very similar results were obtained previously [22,23] using slightly different O–Ar potential parameters (Table 6.1, set 2). The computed isotherm obtained through the use of the Full-Scale (FS) potential (Table 6.1, parameters set 3) is also shown in Figure 6.1. It also agrees quite well with experiments.

6.3.2   Adsorption of methane in $AlPO_4$-5

This system displays an interesting kink (or "step") in the experimental adsorption isotherm [27] at low temperatures, instead of the usual smooth Langmuir (or "type I") isotherm. This step is akin to a phase transition of the confined methane fluid and has first been reproduced by Boutin et al.

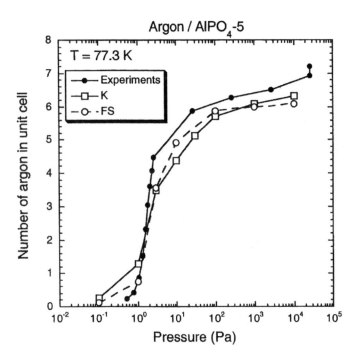

Figure 6.1   Adsorption isotherms of argon in AlPO$_4$-5.

[22,23] through GCMC simulations, using a PN-type potential function. In the first instance, it was thought that a full-scale potential form was presumably needed in order to reproduce such subtle effects as a kink in the isotherm. This was based on the fact that these authors could not find the step using a Kiselev-like potential (with the parameters set 6 given in Table 6.1).

Later, Maris et al. [24] found that this step could be reproduced by using a simple Kiselev-like potential. This has been checked here by performing GCMC simulations, using exactly the same potential parameters (Table 6.1, set 4). As shown in Figure 6.2, the step is indeed obtained in such conditions. The hysteresis observed in Ref. [24] is presumably due to the poor convergence of the simulations in the transition region. A long enough simulation ($\geq 10^8$ steps) leads to the single adsorption–desorption branch shown in Figure 6.2.

It is worth mentioning that Maris et al. [24] claimed that the step could actually be obtained by using the Kiselev-type potential parameters used in Refs. [22,23]. We have fully revisited this point here and we are unable to reproduce their results. We find that the use of the parameters set 6 (Table 6.1) does not lead to a step in the computed isotherm. Different grid spacings in the guest–host potential have been tested and no significant change was found in the resulting form of the isotherms. The same was true when using a combination of Maris et al. and our parameters (set 5, Table 6.1). We thus conclude that, within the Kiselev scheme, slight changes in the methane–oxygen parameters (5% change or less), can transform the type I into a stepped

Table 6.1 Parameters sets of the guest–guest and guest–host potentials used in this work

Set	Guest-Guest		Guest-Host	
1	Ar–Ar	LJ 12–6 [3] $\varepsilon/k_B = (120.0\text{K}, \sigma = 3.405\ \text{Å}$	Ar-O	LJ 12–6 from [24] $\varepsilon/k_B = (86.893\ \text{K}, \sigma = 3.4375\ \text{Å}$
2 [22,23]			Ar-O	LJ 12–6 [26] $\varepsilon/k_B = 124.0\text{K}, \sigma = 3.03\text{Å}$
3 [22,23]			Ar-O, Ar-Al, Ar-P	Full scale
4 [24]**	CH_4–CH_4	LJ 12–6 [24] $\varepsilon/k_B = 148.0\ \text{K}, \sigma = 3.73\ \text{Å}$	CH_4-O [24]	LJ 12–6 [24] $\varepsilon/k_B = (96.5\ \text{K}, \sigma = 3.6\text{Å}$
5			CH_4-O	LJ 12–6 [22,23] $\varepsilon/k_B = (100.0\ \text{K}, \sigma = 3.233\ \text{Å}$
6 [22,23]		LJ 12–6 [36] $\varepsilon/k_B = (148.2\ \text{K}, \sigma = 3.817\ \text{Å}$	CH_4-O	LJ 12–6 [22,23] $\varepsilon/k_B = (100.0\ \text{K}, \sigma = 3.233\ \text{Å}$
7**			CH_4-O	LJ 12–6 [24] $\varepsilon/k_B = 96.5\ \text{K}, \sigma = 3.6\ \text{Å}$
8**			CH_4-O, CH_4-Al, CH_4-P	Full scale [22,23]
9**		LJ 20–6 [37] $\varepsilon/k_B = (148.0\ \text{K}, \sigma = 3.73\ \text{Å}$	CH_4-O	LJ 12–6 [24] $\varepsilon/k_B = 96.5\ \text{K}, \sigma = 3.6\ \text{Å}$
10 [22,23]**			CH_4-O, CH_4-Al, CH_4-P	Full scale [22,23]

**Existence of a step in the computed isotherm.

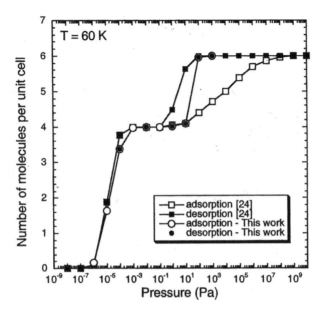

Figure 6.2 Calculated adsorption and desorption isotherms of methane in AlPO$_4$-5 at 60 K.

isotherm displaying the known phase transition. Adsorption isotherms obtained using different parameter sets are shown in Figure 6.3.

In Figure 6.4 is shown the evolution of the pressure at which the step is found in the isotherm, versus temperature. The data obtained with the use of the parameters set 10 in a full-scale potential agree extremely well with experiments. This is not the case with the Kiselev-type potential (parameters set 4) although the slope of the transition pressure is well reproduced.

One important point raised by these studies is the sensitivity of the adsorption data to very small changes in the guest–host steric potential parameters (or to the details of the zeolite structure). This will be a somewhat general conclusion of this survey. Such sensitivities have also been observed in other systems such as pentane/ferrierite [28] and *p*-xylene/silicalite [29], and could partly be attributed to the rigid framework assumption made in the simulations.

6.3.3 Adsorption of xylene isomers in faujasite

Lachet et al. [5,10,30,31] have studied in some details, the adsorption of *p*-xylene and *m*-xylene in several X and Y faujasite zeolites. They were able to reproduce fairly well, the equilibrium adsorption properties using a guest–host potential derived from the PN scheme (Figure 6.5). The transfer-ability of the potential function, from one system to another, has been tested here for these two isomers in NaY and KY faujasites. The potential parameters

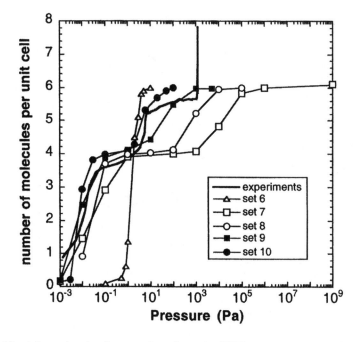

Figure 6.3 Adsorption isotherms of methane in AlPO$_4$-5 at 77.3 K calculated using different potential parameter sets compared to experiments.

were fitted to the experimental data for the m-xylene/NaY system. As shown in Table 6.2, the maximum loading in the isotherm is rather well reproduced in all three other systems, without readjusting the potential parameters. The same calculation has been performed with a Kiselev-type potential, which was obtained by fitting the Lennard–Jones oxygen–xylene atoms parameters in order to reproduce the experimental adsorption isotherm of m-xylene in NaY. As seen in Table 6.2, attempts to transfer this latter potential function to the other xylene/faujasite systems without further readjustment clearly break down.

6.4 Conclusions

Grand Canonical Monte Carlo simulation (using statistical biasing for studying large anisotropic molecules such as xylene isomers), together with an appropriate guest–host forcefield (the Kiselev potential in the simplest cases, a full scale potential in the more complex cases), may provide a reasonably accurate prediction of single component as well as binary mixture adsorption data [5,7].

However, in spite of their impressive results to date on a variety of systems, further progress is still needed in order to reach an acceptable accuracy in the simulations on certain types of systems. Guest–host systems in which the adsorbate molecules fit tightly in the zeolite pores provide

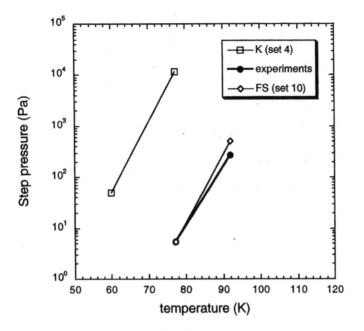

Figure 6.4 Evolution of the pressure at which the step is found in the isotherm versus temperature.

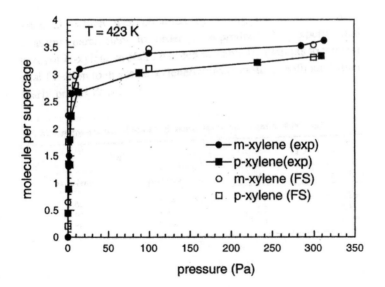

Figure 6.5 Adsorption isotherms of *meta* and *para*-xylene isomers in NaY faujasite.

Table 6.2   Maximum number of adsorbed molecules in the faujasite supercage

Zeolite Isomer	NaY		KY	
	m-xylene	*p*-xylene	*m*-xylene	*p*-xylene
Experiments	3.62	3.34	2.95	3.10
Full scale	3.54	3.31	2.82	3.12
Kiselev-like	3.60	3.05	1.88	2.16

extremely interesting and demanding test cases for the simulation models. The observed sensitivity to small changes in the Kiselev potential terms (or in the zeolite structure) is deemed to be largely unnatural and it illustrates the need to improve the potential models for these (and other) systems. Furthermore, it is in problems such as these that the inclusion of host structure flexibility is probably essential.

We suggest here some further developments in the method used for modeling adsorption of complex mixtures (such as water+hydrocarbons) in cationic zeolites. In these cases, the electrostatic and polarization terms will become dominant and we believe that computing higher order terms of the polarization energy in a self-consistent manner will become necessary. Along these lines, Smirnov [32] has suggested an interesting way of treating molecular partial charges redistribution during adsorption of a highly polar molecule using the electron equalization method.

The guest–guest interaction potential is not the dominant term in the total Hamiltonian of the system, but at high loadings (where most of the interesting features are obtained), details of the intermolecular potential may become crucial. We suggest revisiting the intermolecular forcefields derived from bulk properties, in the manner described by Bayly et al. [33], Kranias et al. [34] and Delhommelle et al. [35]. The main point raised by these authors is that the use of partial charges located only at the atomic sites cannot lead to a good description of the electrostatic potential of the molecule. Using additional electrostatic centers of forces, calculating the partial charges by fitting the *ab initio* electrostatic potential, together with a stabilisation process in order to deal with ill-defined charges in the fitting procedure, leads to very accurate and transferable electrostatic potential terms. This strategy then leaves only two "effective" terms (the repulsive and dispersive ones) that contain adjustable parameters. It has been successful in obtaining transferable potential functions for bulk fluids simulations [35]. The same strategy should now be tested in the case of adsorption. Finally, in view of the unnatural sensitivity of adsorption data to small changes of the potential parameters, the methods used for computing framework partial charges should also be revisited, in order to test the accuracy of the electrostatic field created by the inorganic material at each point in the porous geometry.

Refinement of the simulation models along the lines described above is expected in the near future. They should lead to improved direct predictions of binary mixture adsorption (and presumably transport) properties, and may then help in the rational design of adsorbents.

Acknowledgments

This work was supported by the United States Department of Energy under grant no. DE-FG03-96ER14672, by the French Ministry of Education and the CNRS.

References

1. Shah, R., Payne, M.C., Lee, M.-H., and Gale, J.D. (1996) "Understanding the catalytic behavior of zeolites: a first-principles study of the adsorption of methanol," *Science* 271, 1395.
2. Haase, F., Sauer, J., and Hutter, J. (1997) "Ab initio molecular dynamics simulation of methanol adsorbed in chabazite," *Chem. Phys. Lett.* 266, 397.
3. Schenk, M., Vidal, S.L., Vlugt, T.J.H., Smit," B., and Krishna, R. (2001) "Separation of alkane isomers by exploiting entropy effects during adsorption on silicalite-1: a configurational-bias Monte Carlo simulation study," *Langmuir* 17, 1558–1570.
4. Auerbach, S.M., Bull, L.M., Henson, N.J., Metiu, H.I., and Cheetham, A.K. (1996) "Behavior of benzene in Na-X and Na-Y zeolites: comparative study by ^{2}H NMR and molecular mechanics," *J. Phys. Chem.* 100, 5923–5930.
5. Lachet, V., Buttefey, S., Boutin, A., and Fuchs, A.H. (2001) "Molecular simulation of adsorption equilibria of xylene isomer mixtures in faujasite zeolites. A study of the cation exchange effect on adsorption selectivity," *Phys. Chem. Chem. Phys.* 3, 80–86.
6. Mellot, C.F. and Cheetham, A.K. (1999) "Energetics and structures of fluoro- and chlorofluorocarbons in zeolites: force field development and Monte Carlo simulations," *J. Phys. Chem. B* 103, 3864–3868.
7. Fuchs, A.H. and Cheetham, A.K. (2001) "Adsorption of guest molecules in zeolitic materials: computational aspects," *J. Phys. Chem. B* 105, 7375–7383.
8. Nicholson, D. and Parsonage, N.G. (1982) Computer Simulation and the Statistical Mechanics of Adsorption (Academic Press, New York).
9. Smit, B. and Frenkel, D. (1996) Understanding Molecular Simulation (Academic Press, London).
10. Lachet, V., Boutin, A., Tavitian, B., and Fuchs, A.H. (1998) "Computational study of *p*-xylene/*m*-xylene mixtures adsorbed in NaY zeolite," *J. Phys. Chem. B* 102, 9224–9233.
11. Kiselev, A.V., Lopatkin, A., and Schulga, S.S. (1985) "Molecular statistical calculation of gas adsorption by silicalite," *Zeolites* 5, 261–267.
12. Pellenq, R.J.-M. and Nicholson, D. (1993) "Two body and many-body interactions for argon adsorbed in silicalite zeolites," Proceedings of the Fourth International Conference on Fundamentals of Adsorption, Kyoto (Kodansha, Kyoto), pp. 515–522.
13. Pellenq, R.J.-M. and Nicholson, D. (1994) "Intermolecular potential function for the physical adsorption of rare gases in silicalite-1," *J. Phys. Chem.* 98, 13339–13349.

14. Pellenq, R.J.-M. and Nicholson, D. (1994) "A simple method for calculating dispersion coefficients for isolated and condensed-phase species," *Mol. Phys.* 95, 549–570.

15. Pellenq, R.J.-M., Pellegatti, A., Nicholson, D., and Minot, C. (1995) "Adsorption of argon in silicalite. A semi empirical quantum mechanical study of the repulsive interaction," *J. Phys. Chem.* 99, 10175.

16. Pellenq, R.J.-M. and Nicholson, D. (1995) "Grand ensemble simulation of simple molecules adsorbed in silicalite-1 zeolite," *Langmuir* 2, 1626–1635.

17. Fernandez-Alonso, F., Pellenq, R.J.-M., and Nicholson, D. (1996) "The role of three-body interactions in the adsorption of argon in silicalite-1," *Mol. Phys.* 86, 1021–1030.

18. Nicholson, D. (1996) "Using computer simulation to study the properties of molecules in micropores," *J. Chem. Soc. Faraday Trans.* 92, 1.

19. Nicholson, D. and Pellenq, R.J.-M. (1998) "Adsorption in zeolites: intermolecular interactions and computer simulation," *Adv. Coll. Interf. Sci.* 76, 179–202.

20. Nicholson, D., Boutin, A., and Pellenq, R.J.-M. (1996) "Intermolecular potential functions for adsorption in zeolites: state of the art and effective models," *Mol. Sim.* 17, 217–238.

21. Pellenq, R.J.-M., Tavitian, B., Espinat, D. and Fuchs, A.H. (1996) "Grand canonical Monte-Carlo simulation of adsorption of polar and non-polar molecules in NaY zeolite," *Langmuir* 12, 4768.

22. Lachet, V., Boutin, A., Pellenq, R.J.-M., Nicholson, D., and Fuchs, A.H. (1996) "Molecular simulation study of the structural rearrangement of methane adsorbed in aluminophosphate AlPO$_4$-5," *J. Phys. Chem.* 100, 9006–9013.

23. Boutin, A., Pellenq, R.J.-M., and Nicholson, D. (1994) "Molecular simulation of the stepped adsorption isotherm of methane in AlPO$_4$-5," *Chem. Phys. Lett.* 219, 484–490.

24. Maris, T., Vlugt, T.J.H., and Smit, B. (1998) "Simulation of alkane adsorption in the aluminophosphate molecular sieve AlPO$_4$-5," *J. Phys. Chem. B* 102, 7183–7189.

25. Vlugt, T.J.H., Krishna, R., and Smit, B. (1999) "Molecular simulations of adsorption isotherms for linear and branched alkanes and their mixtures in silicalite," *J. Phys. Chem. B* 103, 1102–1118.

26. Cracknell, R.F. and Gubbins, K.E. (1993) "Molecular simulation of adsorption and diffusion in VPI-5 and other aluminophosphates," *Langmuir* 9, 824–830.

27. Martin, C., Tosi-Pellenq, N., Patarin, J., and Coulomb, J.-P. (1998) "Sorption properties of AlPO$_4$-5 and SAPO-5 zeolite-like materials," *Langmuir* 14, 1774–1778.

28. van Well, W.J.M., Cottin, X., Smit, B., van Hooff, J.H.C., and van Santen, R.A. (1998) "Chain length effects of linear alkanes in zeolite ferrierite. 2. Molecular simulations," *J. Phys. Chem. B* 102, 3952–3958.

29. Cheetham, A.K. and Bull, L.M. (1992) "The structure and dynamics of adsorbed molecules in microporous solids; a comparison between experiments and computer simulations," *Catalysis Lett.* 13, 267–276.

30. Lachet, V., Boutin, A., Tavitian, B., and Fuchs, A.H. (1997) "Grand canonical Monte Carlo simulations of adsorption of mixtures of xylene molecules in faujasite zeolites," *Faraday Discuss.* 106, 307–323.

31. Lachet, V., Boutin, A., Tavitian, B., and Fuchs, A.H. (1999) "Molecular simulation of *p*-xylene and *m*-xylene adsorption in Y zeolites. Single components and binary mixtures study," *Langmuir* 15, 8678–8685.

32. Smirnov, K.S. and Thibault-Starzyk, F. (1999) "Confinement of acetonitrile molecules in mordenite, a computer modeling study," *J. Phys. Chem. B* 103, 8595–8601.

33. Bayly, C.L., Cieplak, P., Cornell, W.D., and Kollman, P.A. (1993) "A well-behaved electrostatic potential based method using charge restraints for deriving atomic charges: the RESP model," *J. Phys. Chem.* 97, 10269.

34. Kranias, S., Boutin, A., Lévy, B., Ridard, J., Fuchs, A.H., and Cheetham, A.K. (2001) "Accurate effective charges and optimized potential for molecular simulation of ethene and some chlorocarbons," *Phys. Chem. Chem. Phys.*, submitted.

35. Delhommelle, J., Tschirwitz, C., Ungerer, P., Granucci, G., Millie, P., Pattou, D., and Fuchs, A.H. (2000) "Derivation of an optimized potential model for phase equilibria (OPPE) for sulfides and thiols," *J. Phys. Chem. B* 104, 4745–4753.

36. Bojan, M.J., Vernov, A.V., and Steele, W.A. (1992) "Simulation studies of adsorption in rough-walled cylindrical pores," *Langmuir* 8, 901–908.

37. Matthews, G.P. and Smith, E.B. (1976) "An intermolecular pair potential energy function for methane," *Mol. Phys.* 32, 1719–1729.

chapter seven

Molecular dynamics simulations for 1:1 solvent primitive model electrolyte solutions

S.-H. Suh*
J.-W. Park
K.-R. Ha
Keimyung University

S.-C. Kim
Andong National University

James M.D. Macelroy
University College Dublin

Contents

* Corresponding author.
Reprint from *Molecular Simulation*, 27: 5–6, 2001. http://www.tandf.co.uk

7.1   Introduction

This chapter, dedicated to Dr David Nicholson, is concerned with molecular dynamics (MD) studies to investigate the structural and transport properties of electrolyte solutions. Among the various theoretical and simulation studies for electrolyte solutions, one of the simplest but most commonly used model systems in the bulk and at interfaces is the so-called "primitive model electrolyte solution" [1]. Within the framework of the primitive model (PM) of electrolyte solutions, the charged hard-sphere ions are immersed in a continuum solvent represented only as a uniform medium of fixed relative permittivity (or dielectric constant). A major drawback in this approach is the use of the solvent continuum assumption in which the solvent structural effects are totally ignored.

In many situations a more detailed representation of the solvent molecules seems necessary and the simplest possible model for the solvent, which retains particulate structure, is known as the solvent primitive model (SPM). In this simple model, which is essentially an extension of the PM to multicomponent form, the solvent particles are treated as neutral hard-spheres of finite size. Although the SPM is clearly an oversimplification in its description of the solvent molecules, an important improvement over PM electrolytes has been observed particularly for systems of high electrolyte concentration where the exclusion packing effects and the short-ranged repulsive interactions are increasingly significant.

Davis and his co-workers [2–4] have successfully applied the SPM electrolytes in their studies of the thermodynamic and structural properties of electric double layers and the effects of solvent exclusion on the force between charged surfaces in electrolyte solutions. In their Monte Carlo studies [3], it was observed that the finite size of the solvent particles resulted in highly ordered layering of ions, which was not captured in the PM electrical double layer. Forciniti and Hall [5] have investigated the equilibrium structure and the thermodynamics of SPM electrolytes, ranging from restricted model electrolytes of the same size to highly asymmetric electrolytes of different sizes, using the hypernetted chain approximation. They found a rather complex but strong correlation between nonelectrostatic and electrostatic contributions to the free energy.

More recently, molecular simulations using both the canonical [6,7] and the grand canonical [8] Monte Carlo (MC) methods have been employed to evaluate the equilibrium thermodynamics and related configurational parameters for SPM electrolytes. In the canonical MC studies reported by Vlachy et al. [6,7], the radial distribution functions were calculated as functions of the neutral solvent concentration and the counterion valency. Evidence of the depletion interaction effect was clearly displayed in the resulting pair correlation functions for highly asymmetric SPM electrolytes, indicating that the addition of a neutral species leads to a gradual change from repulsion to attraction in the qualitative nature of the interactions between similarly charged ions. Similar observations for PM electrolyte solutions [9,10] also suggest that the attraction between like charged macroions is possible if multivalent counterions are present in the solution, in which the valency of the counterions plays an important role in shaping the net interaction between macroions.

Almost all simulations for both PM and SPM electrolyte solutions have been carried out using the MC method. This is mainly due to one principal technical difficulty, namely the "discontinuous" nature of hard-core repulsion combined with "continuous" soft interactions which cannot be handled properly using traditional MD methods [11]. A few implementations [12–14] have been made to extend the MD method of systems of hard cores with soft potentials. A different MD algorithm [15], referred to as the collision Verlet method, was recently introduced and is based on an extension of the general potential splitting formalism. It is interesting to note that this algorithm is nearly identical to our algorithm [14] except that our momenta are defined only at mid time step and a leap-frog formulation is employed.

In the present chapter, we report MD simulation results for the system of symmetric 1:1 SPM electrolytes. In the MD method, the time-dependent transport properties, which cannot be measured by the MC method, are determined by monitoring the actual molecular trajectories as a function of time. In "Model and computations" we describe the interaction model potential and simulation parameters investigated in this work. A brief description of our MD computational techniques is also included. In "Results and discussion," we present the thermodynamics and transport properties obtained from the MD simulations including the collision frequencies, the self-diffusion coefficients, and the velocity and the force autocorrelation functions (FACFs). We also discuss in this section a cluster analysis for the mean cluster size. These simulation results for the cluster and dynamic properties are of particular interest because they can provide specific details of ion cluster formation. Our MD simulation studies can also yield insights into the interplay between short-ranged repulsive and long-ranged attractive interactions.

7.2 Model and computations

In the SPM electrolyte system, the solution is modeled as a mixture of charged ions (solute) and uncharged hard-spheres (solvent) with particle diameter σ immersed in a dielectric continuum ε. For solute/solvent and solvent/solvent interactions, the pair potential between particles i and j is defined as

$$u_{ij}(r) = \begin{cases} \infty & \text{if} \quad r \leq \sigma_{ij} \\ 0 & \text{if} \quad r > \sigma_{ij} \end{cases} \tag{7.1}$$

and, for solute/solute interactions

$$u_{ij}(r) = \begin{cases} \infty & \text{if} \quad r \leq \sigma_{ij} \\ \frac{z_i z_j e^2}{\varepsilon r} & \text{if} \quad r > \sigma_{ij} \end{cases} \tag{7.2}$$

where z_i and z_j are the valences of the ions, e is the charge of the electron, and the additive hard-sphere contact diameter is given by $\sigma_{ij} = (\sigma_i + \sigma_j)/2$.

For the simulations investigated in this work, the uniform dielectric constant ε was chosen to be 78.365 corresponding to water at a room temperature of 298.16 K. Both the positive and negative ions have the same diameter ($\sigma_+ = \sigma_-$) of 4.25 Å and the same charge valency of 1. The diameter of the hard-sphere solvent particles (σ_0) is taken in three separate case studies to be $\sigma_0 = \sigma_+$, $\sigma_0 = 2\sigma_+$, and $\sigma_0 = 5\sigma_+$. All particles including the neutral hard-spheres have the same mass of 100 a.m.u. These parameters are chosen to allow comparisons with previous computational and theoretical studies reported in the literature.

The MD computations were carried out using the minimum image (MI) boundary condition to approximate an infinite system. The long-ranged interaction in the ionic system gives an internal configurational energy that converges slowly with increasing the system size. This is particularly true for higher concentrations and for higher charged systems. For Coulombic plasma systems [16], it has been found that the MI method is sufficiently accurate if the magnitude of a dimensionless parameter,

$$\gamma = \left(\frac{2\pi N}{3V}\right)^{1/3} \frac{(|z_i| + |z_j|)^2 e^2}{\varepsilon k T}$$ (7.3)

is below 10. In our MD simulations a total number of 200 ions ($N_+ = N_- = 100$) was used, and typical values for the parameter condition in Equation 7.3 were less than 1.0. By using a sufficiently large system size, the MI method generates the same accuracy as the Ewald summation method within an acceptable error limit. We observed from a few selected MD runs that a less than 1% relative difference for the configurational energy calculations was achieved in the numerical uncertainty between the MI and the Ewald methods.

The PM or SPM electrolyte system, consisting of a hard-core repulsion with a continuous attractive interaction, gives rise to methodological problems in the MD simulation. Computational approaches in the trajectory calculations are totally different for the discontinuous and the continuous MD methods. Two distinct algorithms were combined within the same MD program by returning to the hybrid method of the "step-by-step" approach described elsewhere [14].

In our MD method the first step is identical to the procedure employed with a continuous potential. The system trajectories are advanced from the current positions to the next positions only under the influence of continuous forces without imposing the hard-core constraints. The next step is then to check whether or not the pair distances are closer than a hard-sphere collision diameter, and, in this step, the particle velocities are assumed to be constant. The algebraic equations of colliding hard-spheres are used to evaluate the collision time between all possible colliding pairs and the resulting configuration is resolved for the overlapping pairs. For computational efficiency, it is appropriate to eliminate any redundant calculations and this was done by constructing a collider table to speed up the search routine.

The equations of motion were integrated using the leap-frog version of the Verlet algorithm with a time step interval of 10^{-14} s. The velocities were scaled at each time step to maintain constant temperature in the manner described by Berendsen et al. [17]. In addition, the starting configurations were generated by randomly inserting particles to assist in the equilibration of the system. Configurations were initially equilibrated for 30,000–50,000 time steps and the final statistics were obtained over $1 \times 10^7 - 2 \times 10^7$ time steps depending on the total number of particles involved.

The MD algorithm implemented in this work has been tested in a number of ways. When the solute ionic charges were assigned to a value of zero, our simulation data faithfully reproduced the pure hard-sphere results. The results obtained from our MD simulations for PM and SPM electrolytes were also compared with previous MC and MD calculations. Good agreement with simulation data reported in the literature again confirmed the quality of our MD algorithm. All simulation runs were performed on the HPC320 of the parallel computing machine at KISTI, Korea. Extensive use was made of optimization and parallelization techniques. About 40 h CPU times were taken in production runs for approximately 2000 particles and 10 million time steps.

7.3 Results and discussion

The thermodynamic and transport properties of 1:1 SPM electrolyte solutions obtained from our MD simulations are presented in Table 7.1. In this table, $\eta_0(= \pi/6\rho_0\sigma_0^3)$ represents the packing fraction of neutral hard-spheres with particle diameter σ_0. For the SPM state conditions, three sets of simulations were performed for $\sigma_0 = \sigma_+$, $\sigma_0 = 2\sigma_+$, and $\sigma_0 = 5\sigma_+$. The PM state point is equivalent to setting $\eta_0 = 0$ in the SPM model. We also report the self-diffusion coefficients and the collision frequencies of both solute ions and solvent hardspheres in the last four columns, respectively.

For the SPM electrolytes the excess internal energy can be written as

$$\frac{U}{N_t kT} = \frac{2\pi\rho_t}{kt} \sum_\alpha \sum_\beta x_\alpha X_B \int_{\sigma_{\alpha\beta}}^\infty u_{\alpha\beta}(r) g_{\alpha\beta}(r) r^2 dr \tag{7.4}$$

and, the virial expression for the osmotic pressure is

$$\frac{PV}{N_t kT} = 1 + \frac{U}{3N_t kT} + \frac{2\pi\rho_t}{3} \sum_\alpha \sum_\beta X_\alpha X_\beta \sigma_{\alpha\beta}^3 g_{\alpha\beta}(\sigma_{\alpha\beta}) \tag{7.5}$$

where X_i is the mole fraction of component i and $g_{ij}(\sigma_{ij})$ is the contact value of the radial distribution function between component i and j at separation distance σ_{ij}.

For ionic solutions the radial distribution function between unlike pairs changes rapidly near the contact point, and, in this case, the extrapolation

Table 7.1 System Characteristics and MD Results for 1:1 SPM Electrolyte Solutions

M_+, M_- (mol/l)	η_0	N_0	$-U/N_i kT (10^{-1})$	$PV/N_i kT$	D_+, D_- (10^{-4} cm²/s)	D_0 (10^{-4} cm²/s)	ω_+, ω_- (10^{11} s⁻¹)	ω_0 (10^{11} s⁻¹)
$\sigma_0 = \sigma_+$								
0.1	0.0		2.7073 (0.0167)	0.9442 (0.0068)	79.614		0.4304	
	0.01	413	0.8909 (0.0064)	1.0366 (0.0068)	36.916	48.933	0.9640	0.7662
	0.02	826	0.5300 (0.0041)	1.0911 (0.0064)	23.253	28.629	1.5145	1.3261
	0.03	1239	0.3823 (0.0024)	1.1417 (0.0068)	16.906	19.992	2.1156	1.9071
2.0	0.0		6.5890 (0.0261)	1.3522 (0.0772)	5.3036		7.1183	
	0.1	207	3.3261 (0.0131)	2.2959 (0.0825)	2.3418	2.5595	18.181	17.123
	0.2	413	2.2632 (0.0086)	3.8610 (0.0978)	1.1752	1.2474	37.488	36.461
	0.3	620	1.7301 (0.0066)	6.7819 (0.1226)	0.5301	0.5479	73.991	72.922
$\sigma_0 = 2\sigma_+$								
0.1	0.1	516	0.7779 (0.0065)	1.4853 (0.0514)	17.445	10.327	2.2655	4.0071
	0.2	1033	0.4697 (0.0038)	2.3805 (0.0654)	8.0253	4.5219	5.0961	9.9440
	0.3	1549	0.3335 (0.0029)	3.9551 (0.0788)	4.1647	2.3310	9.6972	20.372
	0.4	2066	0.2656 (0.0021)	6.9428 (0.0984)	2.1820	1.0051	17.736	40.060
2.0	0.1	26	5.9853 (0.0230)	2.0197 (0.1196)	3.5845	1.9108	11.171	23.561
	0.2	52	5.4956 (0.0209)	3.2396 (0.1783)	2.2591	1.1405	17.674	39.481
	0.3	77	5.1526 (0.0194)	5.4229 (0.2469)	1.3119	0.0573	28.164	66.824
$\sigma_0 = 5\sigma_+$								
0.1	0.1	33	2.3917 (0.0207)	1.1995 (0.0929)	39.894	9.9572	0.9017	3.8515
	0.2	66	2.1739 (0.0173)	1.6907 (0.1502)	22.417	6.0433	1.5906	6.7465
	0.3	99	1.9872 (0.0162)	2.6141 (0.2292)	14.245	3.5440	2.5284	11.658
	0.4	132	1.8691 (0.0127)	4.4197 (0.3376)	8.9152	1.6312	4.0164	20.709
0.2	0.1	2	6.7413 (0.0250)	1.8373 (0.1476)	3.9751	0.4345	9.4528	79.325
	0.2	4	6.9075 (0.0252)	2.6847 (0.2294)	2.7904	0.3109	13.054	114.07
	0.3	5	7.0010 (0.0254)	3.3601 (0.2901)	2.2938	0.2503	15.635	141.25
	0.4	7	7.2131 (0.0250)	5.8603 (0.4402)	1.3519	0.1297	24.414	237.96

Note: The values in parenthesis indicate uncertainties in MD simulations

to the contact value may lead to large uncertainties in MC calculations. For this reason the MC results for osmotic pressure coefficients are known to be less certain than those for the configurational energy. In the MD method better statistics can be achieved in the evaluation of the virial contribution to the equation of state for the hard-core system. The hard-core component of the instantaneous pressure can be obtained from averaging over particle collisions, and the hard-sphere collision contributions to the virial term in Equation 7.5 can be directly calculated during the MD simulations as

$$\frac{PV}{N_t kT} = 1 + \frac{U}{3N_t kT} + \frac{2\pi\rho_t}{3}\frac{1}{t}\sum_{t}^{\text{coll}}\frac{2m_i m_j}{m_i + m_j}(\underline{v}_{ij}\cdot\underline{r}_{ij})\Bigg|_{\text{coll}}. \tag{7.6}$$

For the 1:1 SPM electrolyte solutions investigated here, our simulation results for the internal excess energy and the osmotic pressure have been found to be in close agreement with theoretical approximations using the mean spherical approximation [18] and the hypernetted chain theory [5,6]. Of the two approximations, the hypernetted chain predictions are closer to the MD data. With the exception of the last four entries, inspection of Table 7.1 reveals a rise in the excess internal energy upon the addition of neutral hard-spheres. Note that this value is the averaged one over the total number of particles. However, the configurational energy per ion remains almost constant at a given solvent packing fraction, η_0. The values in parentheses for the thermodynamic results reflect the statistical uncertainties estimated in our MD results, i.e., the standard deviation for block averages over 100 time step segments. Larger deviations in the osmotic pressure are measured for the system at high packing conditions due to the relatively frequent hard-sphere collisions.

The velocity autocorrelation function (VACF) can provide useful insights into ion dynamics and transport. The FACF is another important time correlation function. Although not directly related to time-dependent transport coefficients, the FACF has an important place in the theory of single particle motion. The VACF and the FACF are defined as a function of time t, respectively,

$$\text{VACF} = \frac{1}{N}\sum_{i=1}^{N}\langle\underline{v}_i(0)\cdot\underline{v}_i(t)\rangle \tag{7.7}$$

and

$$\text{FACF} = \frac{1}{N}\sum_{i=1}^{N}\langle\underline{F}_i(0)\cdot\underline{F}_i(t)\rangle \tag{7.8}$$

where the symbol $\langle\rangle$ denotes an average over an equilibrium ensemble.

In Figures 7.1 and 7.2 we display the normalized VACFs and FACFs for the two different sets of concentrations, $\rho_+ = 0.1$ M and $\rho_+ = 2.0$ M, at the fixed

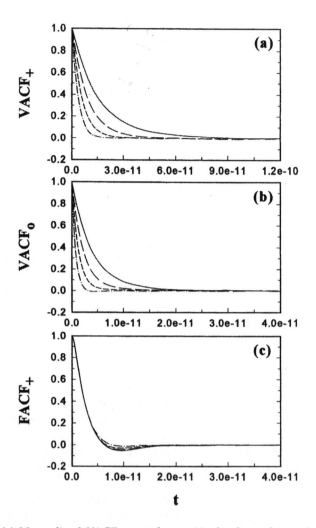

Figure 7.1 (a) Normalized VACFs vs. t for positively charged ions ($\rho_+ = 0.1$), (b) Normalized VACFs vs. t for neutral hard-spheres ($\rho_+ = 0.1$), and (c) Normalized FACFs vs. t for positively charged ions ($\rho_+ = 0.1$). The solid, the long-dashed, the short-dashed, and the chain-dotted curves correspond to $\eta_0 = 0.1$, $\eta_0 = 0.2$, $\eta_0 = 0.3$, and $\eta_0 = 0.4$, respectively.

hardsphere size, $\sigma_0 = 5\,\sigma_+$, respectively, to illustrate the manner in which these functions change with increasing hard-sphere concentration η_0. As shown in these figures, the VACFs of both ions and hard-spheres for $\rho_+ = 2.0$ M decay more rapidly than the corresponding *VACFs* for ρ_+ 0.1 M. The primary mechanism for the decay of the time correlation functions is the hard-sphere collision, in which colliding particles rapidly lose memory of their initial velocities through successive collisions. The $VACF_+$ for the ions exhibits a stronger positive velocity correlation than the $VACF_0$ for the neutral hard-spheres because the Coulombic interaction plays a dominant role in

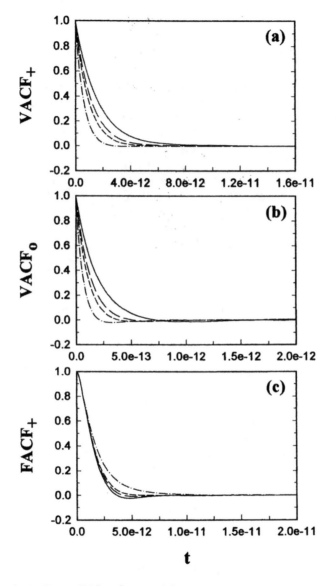

Figure 7.2 As in Figure 7.1 but for $\rho_+ = 2.0$.

determining the particle trajectories of the ions in the low concentration regime. However, at high concentrations, the hard-core repulsive collisions are expected to be the principal contribution to the dynamical properties of these systems. For the high concentration of $\rho_+ = 2.0$ M and $\eta_0 = 0.4$ (shown as the chain-dotted curve in Figure 7.2b) the negative region of the $VACF_0$ indicates that a typical hard-sphere trajectory involves a sequence of back-scattering collisions with its neighboring particles in the first coordination shell.

One of the most interesting features observed in Figures 7.1 and 7.2 is that, while the decay rates of the VACFs and the FACFs are similar at a given ionic concentration, large differences exist between these correlation functions. In comparison with the $VACF_+$, the $FACF_+$ are observed to possess much deeper negative tails particularly for the systems at lower ρ_+ and η_0 values. In the simulation work of Heyes and Sandberg for dense Lennard–Jones systems [9], it was observed that the minimum in the FACF coincides approximately with the cross-over time at which the VACF first changes sign from positive to negative values. One would expect that, in dense systems, the cage effect induced by nearest neighbors dominates the motion of the central particle. This is not the case for dilute electrolyte systems where the electrostatic interaction between unlike pairs of ions tends to create ionic clusters or aggregates. The individual ion particles change their momentum via chattering collisions with their neighbors, while a persistence of velocity is maintained in the original direction of cluster motion. Because the ionic clusters move coherently over a period of time longer than the mean time between ion collisions within the cluster, the VACF has a longer positive correlation than the FACF. The less negative correlation in the FACFs for higher η_0 values, which is opposite to the VACFs, can be explained by the fact that, at higher packing fractions of hard-spheres, the excluded volume effect enhances the formation of larger clusters and this causes restrictions in the coherent motion of ionic clusters.

This last point is clearly illustrated in Figure 7.3 where the mean cluster size S is plotted as a function of cluster cutoff distance, R_{cl}. In our cluster algorithm a pair of dissimilarly charged ions is considered to be within the same cluster if the relative distance between the pair of ions is smaller than a given value of R_{cl}. A similar cluster definition was used in previous simulation studies of 2:2 electrolyte solutions using stochastic Langevin dynamics [20]. The mean cluster size S is obtained from the cluster size distribution using

$$S = \frac{\sum_s s^2 \bar{n}(s)}{\sum_s s \bar{n}(s)} \tag{7.9}$$

where s represents cluster size, and $\bar{n}(s)$ the mean number of clusters of size s. In this work we consider two types of clusters, namely, directly and indirectly bound clusters. Directly bound ions are simply those pairs, that satisfy the geometric cutoff criterion, while indirectly bound ions are connected through intermediate neighboring ions. Such a cluster analysis was implemented using the efficient approach for sampling cluster statistics proposed by Sevick et al. [21].

Figure 7.3 indicates that cluster formation is gradually enhanced with increasing hard-sphere packing fraction η_0. For example, the direct and the indirect mean cluster sizes S within $R_{cl} = 2.0\sigma_+$ are 1.47 and 1.66 at $\rho_+ = 0.1$ and $\eta_0 = 0.1$, and 1.64 and 2.04 at $\rho_+ = 0.1$ and $\eta_0 = 0.4$ as shown in Figure 7.3a; the corresponding S values within $R_{cl} = 1.2\sigma_+$ are 1.95 and 3.38 at $\rho_+ = 2.0$ and $\eta_0 = 0.1$, and 2.50 and 12.34 at $\rho_+ = 2.0$ and $\eta_0 = 0.4$ as shown in Figure 7.3b.

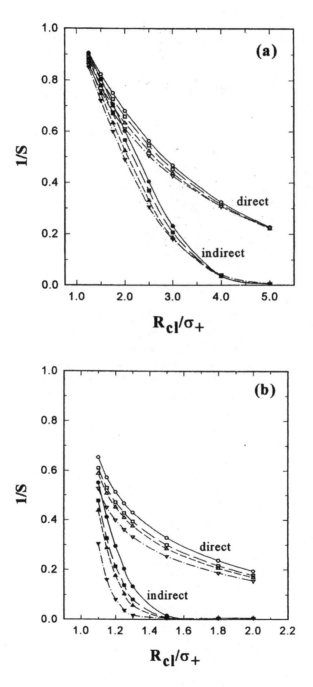

Figure 7.3 The inverse mean cluster size as a function of cluster distance, (a) $\rho_+ =$ 0.1 and (b) $\rho_+ = 2.0$. The symbols of the circle, the square, the upward triangle, and the downward triangle correspond to $\eta_0 = 0.1$, $\eta_0 = 0.2$, $\eta_0 = 0.3$, and $\eta_0 = 0.4$, respectively.

A somewhat large difference between the results for directly and the indirectly bound clusters, particularly in the case of $\rho_+ = 2.0$, suggests the possibility of complex ion cluster formation.

For hard-sphere systems [22], the collision frequencies can be expressed in terms of the contact values of the radial distribution functions

$$\omega_{ij} = \pi \sigma_{ij}^2 \sqrt{\frac{8kT}{\pi \mu_{ij}}} \rho_j g_{ij}(\sigma_{ij}) \tag{7.10}$$

where ω_{ij} is the number of collisions per particle of component i per unit time between particles of component i and j, and $\mu_{ij} (= m_i m_j/(m_i + m_j))$ is the reduced mass. The total collision frequency for component i is simply

$$\omega_i = \sum_{j=1}^{m} \omega_{ij} \tag{7.11}$$

In Figure 7.4 the collision frequencies vs. η_0 determined from the MD simulation are illustrated for solute ions, ω_+ (Figure 7.4a), and solvent hard-spheres, ω_0 (Figure 7.4b), respectively. For the purpose of comparison with the corresponding hard-sphere systems, theoretical predictions for the collision frequencies obtained using Equation 7.10 in conjunction with contact values for the radial distribution functions computed directly from the MD simulations are also shown as the dotted curves in these figures. The MD results are seen to be in excellent agreement with hard-sphere approximations over a wide range of η_0 and σ_0. This suggests that the microscopic dynamics of the SPM electrolyte solutions investigated in this work are very similar to the dynamical processes taking place in neutral hard-sphere mixtures. In this sense the transport properties of 1:1 SPM systems are, at least qualitatively, related to those for hard-sphere fluids.

In previous MD simulations of 1:1 PM electrolyte solutions [13], the Enskog theory of hard-spheres was shown to predict the self-diffusion coefficients reasonably accurately. Such a modified Enskog approximation for PM electrolytes can also be extended to the SPM electrolyte systems. In the extended Enskog theory the self-diffusion coefficient for component i can be expressed in terms of the intradiffusion coefficients for multicomponent mixtures,

$$D_{E,i} = \left(\sum_{j=1}^{m} \frac{\rho_j g_{ij}(\sigma_{ij})}{(\rho D_{ij})_0} \right)^{-1} \tag{7.12}$$

and

$$(\rho D_{ij})_0 = \frac{3}{8\sigma_{ij}^2} \sqrt{\frac{kT}{2\pi \mu_{ij}}} \tag{7.13}$$

where $(\rho D_{ij})_0$ represents the product of the number density and the binary mutual diffusion coefficient in the dilute gas limit of hard-spheres.

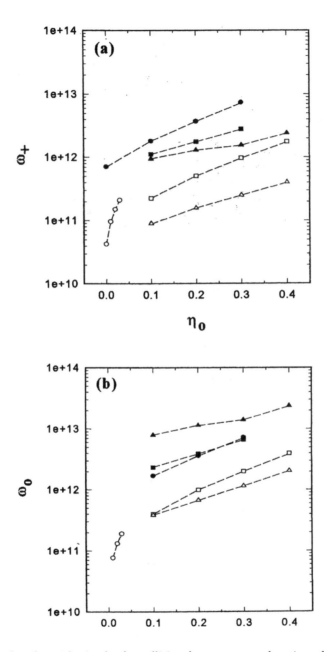

Figure 7.4 Semilogarithmic plot for collision frequency as a function of η_0. (a) positively charged ions and (b) neutral hard-spheres. The symbols of the circle, the square, and the triangle correspond to $\sigma_0 = \sigma_+$, $\sigma_0 = 2\sigma_+$, and $\sigma_0 = 5\sigma_+$, and the open and filled symbols represent $\rho_+ = 0.1$ and $\rho_+ = 2.0$, respectively. The dotted curves are theoretical predictions provided by Equations 7.10 and 7.11 using the MD contact values for the radial distribution functions.

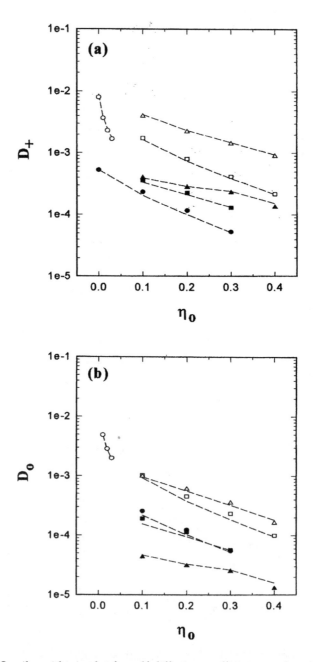

Figure 7.5 Semilogarithmic plot for self-diffusion coefficient as a function of η_0. (a) positively charged ions and (b) neutral hard-spheres. The symbols are the same as in Figure 7.4. The dotted curves are theoretical predictions provided by Equations 7.12 and 7.13 in conjunction with the MD contact values for the radial distribution functions.

In Figure 7.5 the self-diffusion coefficients obtained from the MD simulations are compared with those predicted using the extended Enskog theory in Equations 7.12 and 7.13. The MD data for the diffusivities were calculated from the integration of the corresponding VACF using the Green-Kubo relationship. Again, it is observed that the theoretical predictions and the MD calculations are in good agreement. One should recall however that, while the Enskog theory takes advantage of the simplification that the properties of a dense fluid are primarily determined by the repulsive core of the particle–particle interaction, it does have limitations, particularly a high density. The error involved in using the Enskog theory under these conditions is ascribed to the failure of the molecular-chaos approximation and the deviations are most pronounced when the cage effect is important. For mixtures of 1:3 PM electrolytes [14], it was found that the self-diffusion coefficients of the lower charged electrolytes were close to those for 1:1 PM electrolytes, whereas those of highly charged electrolytes were smaller by a factor of two or three. An interpretation of this observation is that the free motion of highly charged ions is likely to be restricted by the formation of ionic clusters.

7.4 Conclusion

In the present work MD simulations at constant temperature have been carried out to investigate the equilibrium thermodynamic and time-dependent transport properties of 1:1 solvent PM electrolyte solutions. MD results for the excess internal energy and the osmotic pressure are shown to be in good agreement with the mean spherical approximation, and, more precisely, with the hypernetted chain theory. In the lower concentration regime, the electrostatic interaction plays an important role in determining ion trajectories, while the hard-sphere collisions dominate in the higher concentration regime.

Significant differences are also observed between the VACFs and FACFs. The less negative correlation effects displayed by the FACFs at higher hard-sphere packing fractions are related to a restricted coherent motion of the larger ionic colusters, which are formed at these densities. This conclusion is supported by an independent analysis of the direct and indirect bound ion cluster size distributions computed during the MD simulations. Under the conditions employed in this work, excellent agreement is also observed between the MD results and the theoretical predictions for the self-diffusion coefficients and the collision frequencies of both ionic solute and hard-sphere solvent. In this respect our simulation studies strongly suggest that by incorporating the discrete particulate nature of the solvent into models of electrolyte solutions, then the interpretation of the nonequilibrium as well as equilibrium phenomena taking place within such systems should be significantly improved.

Acknowledgments

This work was supported by a grant from KOSEF and the assistance of computing resources from KISTI. JWP is also grateful to this graduate stipend through the BK21 project.

References

1. Durand-Vidal, S., Simonin, J.-P. and Turq, P. (2000) *Electrolytes at Interfaces* (Kluwer Academic Publishers, Dordrecht).
2. Tang, Z., Scriven, L.E. and Davis, H.T. (1992) "A three-component model of the electrical double layer," *J. Chem. Phys.* 97, 494–503.
3. Zhang, L., Davis, H.T. and White, H.S. (1993) "Simulations of solvent effects on confined electrolytes," *J. Chem. Phys.* 98, 5793–5799.
4. Tang, Z., Scriven, L.E. and Davis, H.T. (1994) "Effects of solvent exclusion on the force between charged surfaces in electrolyte solution," *J. Chem. Phys.* 100, 4527–4530.
5. Forciniti, D. and Hall, C.K. (1994) "Structural properties of mixtures of highly asymmetrical electrolytes and uncharged particles using the hypernetted chain approximation," *J. Chem. Phys.* 100, 7553–7566.
6. Rescic, J., Vlachy, V., Bhuiyan, L.B. and Outhwaite, C.W. (1997) "Monte Carlo simulation studies of electrolyte in mixture with a neutral component," *J. Chem. Phys.* 107, 3611–3618.
7. Rescic, J., Vlachy, V., Bhuiyan, L.B. and Outhwaite, C.W. (1998) "Monte Carlo simulations of a mixture of an asymmetric electrolyte and a neutral species," *Mol. Phys.* 95, 233–242.
8. Wu, G.-W., Lee, M. and Chan, K.-Y. (1999) "Grand canonical Monte Carlo simulation of an electrolyte with a solvent primitive model," *Chem. Phys. Lett.* 307, 419–424.
9. Hribar, B. and Vlachy, V. (1997) "Evidence of electrostatic attraction between equally charged macroions induced by divalent counterions," *J. Phys. Chem. B* 101, 3457–3459.
10. Hribar, B. and Vlachy, V. (2000) "Clustering of macroions in solutions of highly asymmetric electrolytes," *Biophys. J.* 78, 64–698.
11. Allen, M.P. and Tildesley, D.J. (1987) *Computer Simulation of Liquids* (Oxford Science Publications).
12. McNeil, W.J. and Madden, W.G. (1982) "A new method for the molecular dynamics simulation of hard core molecules," *J. Chem. Phys.* 76, 6221–6226.
13. Heyes, D.M. (1982) "Molecular dynamics simulations of restricted primitive model 1:1 electrolytes," *Chem. Phys.* 69, 155–163.
14. Suh, S.-H., Mier-y-Teran, L., White, H.S. and Davis, H.T. (1990) "Molecular dynamics study of the primitive model of 1-3 electrolyte solutions," *Chem. Phys.* 142, 203–211.
15. Houndonougbo, Y.A., Laird, B.B. and Leimkuhler, B.J. (2000) "A molecular dynamics algorithm for mixed hard-core/continuous potentials," *Mol. Phys.* 98, 309–316.
16. Brush, S.G., Sahlin, H.L. and Teller, E. (1966) "Monte Carlo study of a one-component plasma. I," *J. Chem. Phys.* 45, 2102–2118.

17. Berendsen, H.J.C., Postma, J.P.M., van Gunsteren, W.F., DiNola, A. and Haak, J.R. (1984) "Molecular dynamics with coupling to an external bath," *J. Chem. Phys.* 81, 3684–3690.

18. Sanchez-Castro, C. and Blum, L. (1989) "Explicit approximation for the unrestricted mean spherical approximation for ionic solutions," *J. Chem. Phys.* 93, 7478–7482.

19. Heyes, D.M. and Sandberg, W.C. (1990) "Microscopic motion of atoms in simple liquids at equilibrium and with shear flow," *Phys. Chem. Liq.* 22, 31–50.

20. Abascal, J.L.F., Bresme, F. and Turq, P. (1994) "The influence of concentration and ionic strength on the cluster structure of highly charged electrolyte solutions," *Mol. Phys.* 81, 143–156.

21. Sevick, E.M., Monson, P.A. and Ottino, J.M. (1987) "Monte Carlo calculations of cluster statistics in continuum models of composite morphology," *J. Chem. Phys.* 88, 1198–1206.

22. McQuarrie, D.A. (1976) *Statistical Mechanics* (Harper and Row, New York).

chapter eight

Computer simulation of isothermal mass transport in graphite slit pores

K. P. Travis[*]
University of Bradford

K. E. Gubbins
North Carolina State University

Contents

[*] Corresponding author.
Reprint from *Molecular Simulation*, 27: 5–6, 2001. http://www.tandf.co.uk

8.1 Introduction

Porous materials are used extensively in the petroleum and chemical process industries as catalysts and adsorbents. Of the various contributions to the flow of fluid through these materials, diffusion is the most important, since more often than not, it is the rate determining process. To facilitate the design of improved catalytic and adsorption processes, a greater understanding of the complexities of diffusional behaviour, particularly at the molecular level, is required. Computer simulation is ideally suited to this goal, providing a direct link between the microscopic properties of molecules and macroscopic properties, which are measured in the laboratory.

The motivation behind our current study stems from the important industrial process by which air is separated into its major components by pressure swing adsorption (PSA). In this diffusion-controlled process, a stream of air is passed through a bed of molecular sieving carbon, an adsorbent containing micropores with a mean width of 0.5 nm.

Oxygen selectivities of between 3 and 30 have been reported, even though the kinetic diameters of oxygen and nitrogen differ by less than 0.03 nm. A precise explanation for these large selectivity values remains elusive, despite considerable research. It is important to be able to identify the key parameters in this diffusion process, and then to find their optimum values in order to maximize the amount of oxygen recovered while maintaining economic viability. Several parameters can influence the transport rates of fluids through adsorbents such as molecular sieving carbon, temperature, pore size, and pore morphology being just a few examples.

The effect of pore width on oxygen selectivity can be probed by experimental methods. Chihara and Suzuki [1] attempted to vary the ratio of diffusivities of oxygen and nitrogen in molecular sieving carbon by adsorption of hydrocarbons followed by heat treatment. They concluded that the absolute diffusivities of oxygen and nitrogen could be decreased by an order of magnitude. However, changing the mean pore width in the adsorbent cannot vary the ratio of their diffusivities. Computer simulation results appear to contradict the finding of Chihara and Suzuki; Seaton et al. [2] studied the separation of oxygen and nitrogen in model graphitic pores. They conducted molecular dynamics simulations of self-diffusion in individual pores, and found that the diffusivities were strongly dependent on the pore width. Using a randomly etched graphite pore model (REGP) they found that the degree of kinetic separation observed experimentally could be reproduced at the level of individual pores. In a more recent publication MacElroy et al. [3] looked at transport diffusion of oxygen and nitrogen in the same model pore system, concluding that pore length was a controlling factor in the separation mechanism. Recently, Travis and Gubbins [4] investigated the role of pore width on transport diffusion of oxygen and nitrogen mixtures flowing through a single slit using non-equilibrium molecular dynamics (NEMD) techniques. No significant differences were found between the component diffusivities except at the lowest pore width studied

(0.8375 nm). At this pore width, nitrogen diffuses faster than oxygen, in contradistinction to the experimental observation.

The model used by Travis and Gubbins contained several approximations, for example, the graphitic adsorbent was modelled as a single, smooth walled slit pore, with no account taken of surface structure or electrostatic effects. However, lack of surface structure and electrostatic effects in the model are thought not to be important at ambient temperatures. Furthermore, the use of a single slit pore model of the adsorbent both aids data analysis and provides results that can be used as input in network models. A key difficulty encountered in our earlier study was the interpretation of the mixture transport coefficients. In this chapter, we address this problem by including simulation results for pure component diffusion at similar conditions to the mixture. We also employ equilibrium molecular dynamics (EMD) and grand canonical Monte Carlo (GCMC) simulations to examine how the various contributions to the diffusion coefficients vary with pore width and temperature.

We have organised the chapter as follows: in Section 8.2, we discuss the transport equations for single micropores. In Section 8.3, we discuss the computer simulation algorithms including the technique of Dual Control Volume Grand Canonical Molecular Dynamics (DCV GCMD), which we have used to obtain most of our diffusion data. In Section 8.4, we discuss the model and simulation details, and in Section 8.5, we present and discuss our results. Finally, in Section 8.6 we present our conclusions.

8.2 Transport in single micropores

The starting point for discussing transport in porous membranes is the Dusty Gas model developed by Mason and co-workers [5,6]. The main assumption in this model is that the solid particles, which comprise the membrane, can be treated as if they were a component in the diffusing mixture. This is justified on the grounds that if the adsorbate gas is at low density, a representative volume element must be large enough to contain several molecular mean free paths in order for the postulate of local thermodynamic equilibrium to hold within the volume element. In this case the volume element will contain some of the membrane particles (the "dust"). A single pure gas flowing through a membrane therefore becomes a binary system. A direct manifestation of this treatment is the presence of both viscous and diffusive terms in the flux expressions describing fluid transport through a membrane. In extreme cases, one of these transport modes will dominate the other. A description of the transport process will then be a furnished by either Fick's law of diffusion, or Poiseuille's law. In membranes with very wide pores, viscous flow can be expected to dominate, while in very narrow pores, diffusion should dominate.

The equations describing the isothermal transport of a multi-component fluid mixture through a membrane can be given either in the Stefan–Maxwell form [7] or, equivalently, in linear irreversible thermodynamic form. The former are the most useful from an engineering point of view, while the latter are more useful in computer simulation studies since the kinetic transport coefficients can be directly related to equilibrium time correlation functions.

The Stefan–Maxwell equations for multi-component fluid flow in a membrane with slit pore geometry (and assuming no viscous separation) are

$$\frac{-1}{k_B T}\left(\frac{\partial \mu_n}{\partial x}\right) = \sum_{m=1}^{K} \frac{\rho_m}{\rho D_{nm}}(u_{nx} - u_{mx}) + \frac{u_{nx}}{D_{nM}} + \frac{B_0}{\eta D_{nM}}\left(\frac{\partial p}{\partial x}\right) \tag{8.1}$$

where μ_n is the chemical potential of fluid species n, T the temperature, ρ the mean fluid density, ρ_m the density of fluid species m, u_n the stream velocity of fluid species n, B_0 is a constant characteristic of the membrane geometry, η is the shear viscosity, p the hydrostatic pressure, while D_{nm} is the Stefan–Maxwell coefficient representing the interdiffusion of fluid species n and m. D_{nM} is the Stefan–Maxwell coefficient representing the diffusion of fluid species n in the membrane denoted by the subscript M.

The linear irreversible thermodynamic expression for the component flux is

$$J_{mx} = -\sum_n L_{mn}\left(\frac{\partial \mu_n}{\partial x}\right) - L_0\left(\frac{\partial p}{\partial x}\right) \tag{8.2}$$

where J_{mx} is the flux of component m in the x-Cartesian direction of a laboratory frame of reference, L_{mn} are the phenomenological transport coefficients and L_0 is a viscous transport coefficient. The phenomenological coefficients, L_{mn}, are related to microscopic properties of the fluid through Green-Kubo type formulae or their equivalent Einstein mean square displacement formulae

$$L_{mn} = \frac{N_m N_n}{2 V K_B T}\int_0^\infty \langle u_m(t)u_n(0)\rangle dt \tag{8.3}$$

$$L_{mn} = \frac{N_m N_n}{4 V K_B T}\lim_{t\to\infty}\frac{d}{dt}\langle [R_m(t)-R_m(0)][R_n(t)-R_n(0)]\rangle \tag{8.4}$$

where R_m is the center-of-mass of fluid component m and N_m is the number of molecules of type m. In the special case of single component fluid transport, these equations become, respectively,

$$L_f = \frac{N^2}{2 V K_B T}\int_0^\infty \langle u(t)u(0)\rangle dt \tag{8.5}$$

$$L_f = \frac{N^2}{4 V k_B T}\lim_{t\to\infty}\frac{d}{dt}\langle [R(t)-R(0)]^2\rangle \tag{8.6}$$

where L_f is the single component transport coefficient.

In the single component case, a simple one to one mapping exists between the phenomenological coefficient L_f and a Stefan–Maxwell diffusion coefficient, $D_{0M}(\equiv D_0)$ which is the limiting case of D_{nM}. This follows from substituting the flux expression ($J = \rho u$) into Equation 8.1, rearranging, and then comparing with Equation 8.2 to give

$$D_0 = \frac{k_B T}{\rho} L_f \tag{8.7}$$

We shall henceforth refer to D_0 as the collective diffusivity. The situation for mixtures is more complicated. Only in the case of a binary fluid mixture can tractable relations be derived. The mapping is then

$$\frac{D_{1M}}{k_B T} = \frac{L_{11}L_{22} - L_X^2}{\rho_1 L_{22} - \rho_2 L_X} \tag{8.8}$$

$$\frac{D_X}{k_B T} = \frac{L_{11}L_{22} - L_X^2}{(e_1 + e_2)L_X} \tag{8.9}$$

where L_X and D_X represent the cross terms L_{12} and D_{12}, which are identical to L_{21} and D_{21} by symmetry. The equation for D_{22} can be obtained by interchanging the indices in Equation 8.8. The inverse relationships can also be written down. These are

$$k_B T L_{11} = \frac{(\rho D_{12} + \rho_1 D_{2M})\rho_1 D_{1M}}{\rho D_{12} + \rho_1 D_{2M} + \rho_2 D_{1M}} \tag{8.10}$$

$$k_B T L_x = \frac{(\rho_1 \rho_2 D_{1M} D_{2M})}{\rho D_{12} + \rho_1 D_{2M} + \rho_2 D_{1M}} \tag{8.11}$$

with the equation for L_{22} being obtained by interchanging indices in Equation 8.10.

The total intrapore flux for a single component fluid flowing through a single slit pore becomes (using Equations 8.2 and 8.7 and $L_0 = \rho B_0 / \eta$),

$$J_x = -\frac{\rho D_0}{k_B T}\left(\frac{\partial \mu}{\partial x}\right) - \frac{\rho B_0}{\eta}\left(\frac{\partial p}{\partial x}\right) \tag{8.12}$$

where B_0 is a geometric factor characteristic of a slit pore geometry and η is the coefficient of shear viscosity. The first term on the right hand side of Equation 8.12 is readily identified as the diffusive contribution to the flux, J_x^D, while the other term is the viscous contribution to the flux, J_x^V. The expression for the total intrapore flux is then, formally,

$$J_x^{tot} = J_x^D + J_x^V \tag{8.13}$$

Using the Gibbs—Duhem equation, the chemical potential gradient appearing in the expression for the diffusive flux can be transformed into a density gradient with the result that

$$J_x^D = -D_0 \left(\frac{d\ln f}{d\ln \rho} \right) \left(\frac{\partial \rho}{\partial x} \right) \tag{8.14}$$

where f is the fugacity of the external gas phase that is in equilibrium with the adsorbate. The equation is now in the more familiar Fickian form. The first quantity in parentheses is a thermodynamic factor which is known as the Darken factor. In words, it is the inverse slope of the adsorption isotherm in logarithmic coordinates. The product of D_0 with the Darken factor is referred to as the transport diffusivity, D_t,

$$D_t = D_0 \left(\frac{d\ln f}{d\ln \rho} \right) \tag{8.15}$$

which is the constant of proportionality between the diffusive flux and the density gradient. The form of Equation 8.15 shows why D_0 is sometimes referred to as the corrected diffusivity; it has to be corrected for the thermodynamic factor [8]. The transport diffusivity can be expected to have a stronger concentration dependence than the collective diffusivity, D_0, as a result of the concentration dependence of the Darken factor. The Darken factor approaches unity in the limit of vanishing density.

The collective diffusivity, D_0, can be shown to consist of a self diffusivity, D_S, and a diffusivity which arises through momentum cross coupling, D_ξ [9]

$$D_0 = D_s + D_\xi \tag{8.16}$$

In the limit of zero loading, the cross coupling diffusivity will vanish and D_0 becomes equal to the self diffusivity, D_S, a single particle property.

The pressure gradient appearing in the expression for the viscous flux can be rewritten as a density gradient with the effect that an effective transport diffusivity, D_t^{eff}, may be defined as

$$J_x^{tot} = -D_t^{eff} \left(\frac{\partial \rho}{\partial x} \right) \tag{8.17}$$

where the effective diffusivity now contains a viscous contribution. The significance of Equation 8.17 will be explained in the next section.

8.3 Calculation of transport properties via computer simulation

In principle, all the transport coefficients appearing in Equation 8.2 could be calculated in a single EMD simulation and used to calculate values of the Stefan–Maxwell diffusion coefficients. Once these quantities are known, together with the shear viscosity, the intrapore fluxes may be predicted for a range of different driving forces. However, in practice, the integrals in Equations 8.3 and 8.5 are notoriously difficult to calculate because the time correlation functions exhibit long time tails and suffer from a poor signal to noise ratio. Recently, a new NEMD method was developed [10–12] which allows the direct simulation of mass transport through membranes under the influence of a chemical potential gradient. This method has been named Dual Control Volume Grand Canonical Molecular Dynamics (DCV GCMD). This type of simulation gives directly the total intrapore fluxes and, as an added bonus, can yield the diffusion coefficients with superior signal to noise compared to the EMD route. With the introduction of more accurate models for membranes and improved intermolecular potentials, DCV GCMD can be expected to be a powerful tool in predicting membrane separation performance.

   In the DCV GCMD technique, a gradient in chemical potential is established by placing two particle reservoirs at either end of a single pore and maintaining them at fixed, but different, chemical potentials. This is achieved by periodically conducting a series of creations and deletions according to the prescription of grand canonical Monte Carlo (GCMC). Examples of the uses of DCV GCMD include: a study of the transport diffusion of methane in graphite [12], diffusion of gases in zeolite frameworks [13], diffusion through polymer membranes [14] and diffusion of a mixture of oxygen and nitrogen through a graphite slit-pore [3,4,15].

   The main drawback of DCV GCMD is the difficulty in extracting the diffusion coefficients in an unambiguous manner. Consider the case of a single component fluid in a slit-pore as an example. DCV GCMD gives the total flux directly. However, the total flux in general consists of a diffusive term and a viscous term (Equation 8.13) and the diffusion coefficient obtained from taking the ratio of the total flux to the chemical potential gradient is only an effective diffusion coefficient. There is no simple way to uncouple the two contributions to the total flux. One solution, which has been tried, is to assume the viscous flux is given by the solution of a Poiseuille flow problem [16]. However, such a method is doomed to failure because the quadratic velocity profile predicted by classical hydrodynamics is not observed in pores of less than about 10 molecular diameters in width [17,18] Travis and Gubbins introduced a more fruitful approach [19]. Their solution was to perform the DCV GCMD experiment first and then, knowing the mean density and total flux, perform a second simulation, but this time of pure Poiseuille flow at the same density and equivalent pressure gradient. This relies on the use of a constant force rather than an actual pressure gradient to drive the flow, so that the density profile remains uniform in the flow direction. Since no diffusion occurs, the integrated flux profile yields directly the viscous contribution to the total DCV GCMD flux. The main drawback to this method is that an extra simulation needs to be performed. In some situations the viscous flux is negligible in comparison to the diffusive flux and the problem no longer arises.

One further problem with DCV GCMD is the difficulty in calculating an accurate value of the local chemical potential. It is often more convenient to calculate the local density than the local chemical potential. The result is that the more fundamental transport coefficient, D_0, cannot be obtained directly from a DCV GCMD simulation. To obtain D_0 indirectly from a DCV GCMD simulation, one must perform a series of GCMC simulations, obtain the adsorption isotherm and then calculate the Darken factor from this. In the limit that $\rho \to 0$, the Darken factor becomes unity and DCV GCMD then gives D_0 directly. However, in this limit, D_0 can be replaced by D_S, the self diffusivity that is more readily obtained by EMD.

The same arguments apply when fluid mixtures are considered: viscous terms. Darken factors, and now, cross coefficients mean that obtaining the Stefan–Maxwell coefficients directly from a DCV GCMD simulation is an arduous task. In order to fully understand the mechanism of diffusion controlled separation processes it is therefore necessary to use a simulation "toolkit" containing in addition to DCV GCMD, EMD, Poiseuille flow NEMD and GCMC methods. In the present work we report simulations employing a range of these techniques. We have chosen to use a simple slit pore representation of the molecular sieving carbon to facilitate interpretation of the data.

8.4 Simulation details

8.4.1 Adsorbate and adsorbent models

We represent the interaction of the adsorbate molecules with the graphite planes in our model by a smooth 10–4–3 potential due to Steele [20]. This potential is a function of the z-co-ordinate only. Neglecting the corrugations in the xy graphite planes is expected to be a reasonable approximation for this study, but might be a serious omission at lower temperatures. The total potential energy function, which takes into account the interactions of both graphite planes in a slit-pore is

$$\Phi_{ic} = 4\pi\varepsilon_{ic}\sigma_{ic}^2 \Delta n_c \left[\frac{1}{5}\left(\frac{\sigma_{ic}}{(H/2-z)} \right)^{10} - \frac{1}{2}\left(\frac{\sigma_{ic}}{(H/2-z)} \right)^4 \right.$$

$$- \frac{\sigma_{ic}^4}{6\Delta(H/2-z+0.61\Delta)^3} + \frac{1}{5}\left(\frac{\sigma_{ic}}{(H/2+z)} \right)^{10} - \frac{1}{2}\left(\frac{\sigma_{ic}}{(H/2+z)} \right)^4 \qquad (8.18)$$

$$\left. - \frac{\sigma_{ic}^4}{6\Delta(H/2+z+0.61\Delta)^3} \right]$$

where Δ is the inner layer spacing in graphite, which is taken to be 0.335 nm, $n_c = 114$ nm^{-3} is the carbon atom number density in graphite, H is the pore width, defined as the distance between the centers-of-mass of the innermost graphite planes, while ε_{ic} and σ_{ic} are the Lennard–Jones parameters appropriate for interactions between a molecular site of species i and a carbon atom.

Oxygen and nitrogen are modelled as two-center Lennard-Jones molecules with fixed bond lengths. Interactions between sites on different fluid molecules are modelled with a truncated and shifted Lennard-Jones 12–6 potential:

$$\Phi_{ij}(r) = \begin{cases} 4\varepsilon_{ij}\left(\frac{\sigma_{ij}}{r_{ij}}\right)^{12} - \left(\frac{\sigma_{ij}}{r_{ij}}\right)^6 - \Phi_{ij}(r_c) & r \le r_c \\ 0 & r > r_c \end{cases} \tag{8.19}$$

In the above equation, r is the scalar interatomic distance between a pair of interacting sites, r_c is the truncation distance, and $\Phi_{ij}(r_c)$ is the value of the potential energy at the point of truncation. Lennard-Jones potential parameters for nitrogen, oxygen, and carbon were taken from the literature [21] and are given in Table 8.1. Parameters appropriate for interactions between chemically different species, for example, between an oxygen atom and a nitrogen atom, are given by the Lorentz-Berthelot mixing rules: $\sigma_{ij} = 1/2(\sigma_i + \sigma_j)$ and $\varepsilon_{ij} = \sqrt{\varepsilon_i \varepsilon_j}$. The bond lengths of the molecules are 0.1097 nm for nitrogen and 0.1169 nm for oxygen. We truncate the Lennard-Jones potential at $r_c = 2.5\sigma_{ij}$ (we do not truncate or shift the Steele 10–4–3 potential). We consider only classical dynamics in constructing our model, quantum effects being unimportant for a system such as ours [22]. No account is taken of the quadrupole for nitrogen molecules. Justification for this approximation comes from simulation studies of nitrogen adsorption in slit pores at ambient temperature, which found that the quadrupole had no significant effect on the results [23].

8.4.2 Dual control volume simulations

The DCV GCMD algorithm for use with multi-component mixtures has been discussed in detail elsewhere [4], so we give only a brief description of its implementation here.

There are several variations of DCV GCMD but essentially the algorithm consists of performing numerous cycles, each of which comprises a molecular dynamics step, in which the trajectories of all fluid molecules are incremented, followed by a series of Grand Canonical Monte Carlo creations and destructions of either species in each of the two control volumes.

Table 8.1 Lennard-Jones potential parameters for oxygen, nitrogen and carbon used in this work

Atom	σ (nm)	$k_B^{-1}(\varepsilon/K)$
Carbon	0.340	28.0
Nitrogen	0.3296	60.39
Oxygen	0.2940	75.49

A creation attempt of a molecule of species γ in control volume c is accepted with a probability,

$$\min[1, \exp[-\beta\Delta v_\gamma + \ln(z_\gamma(c)V(c)/(N_\gamma + 1))]] \tag{8.20}$$

where $V(c)$, N and $z(c)$ are the volume, the current number of molecules in control volume c and the activity in control volume c, respectively, while Δv is the energy change accompanying the insertion of a molecule into the control volume, and $\beta = 1/k_BT$.

Destructions of molecules are accepted with a probability

$$\min[1, \exp[-\beta\Delta v_\gamma + \ln(N_\gamma / (z_\gamma(c)V(c)))]] \tag{8.21}$$

where Δv is now the energy change accompanying the destruction of a molecule from control volume c.

When molecules are created in either control volume, they are assigned thermal components of both translational velocity and angular velocity selected from a Maxwell–Boltzmann distribution. Furthermore, newly created molecules are given an appropriate initial component of streaming velocity to ensure creations are compatible with mass transport (further details are given below).

Control volumes are placed at each end of the slit-pore. Placing the control volumes inside the pore eliminates pore-entrance effects and greatly simplifies the interpretation of our results. We define our co-ordinate system such that the flow is the x-direction and the graphite planes are separated along the z-direction. The volume of the control volumes is taken to be the same as that of the flow region in between them (see Figure 8.1). We note that there is no unambiguous definition of volume in a porous membrane. However for simplicity, we define the volume of the flow region and control volumes to be $V = HL_xL_y$, which contains a certain amount of dead space due

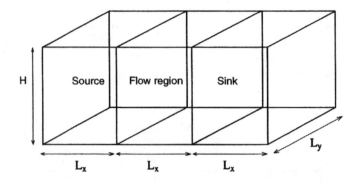

Figure 8.1 Schematic diagram of the simulation cell for DCV GCMD. Flow takes place in the x-direction from the source to the sink control volume.

to the implicit carbon atoms in graphite. The control volumes and the flow region have a length in the x-direction of $L_x = 9.888$ nm, while in the y-direction, $L_y = 9.888$ nm. The total number of molecules in the simulation cell varied between 1200 and 3600 molecules.

The system is kept isothermal via the following scheme: The simulation box is divided into 21 bins of equal volume along the flow direction. The local temperature in each bin is then controlled by means of the Nosé–Hoover thermostat [24,25]. In a molecular dynamics step, a fifth order Gear algorithm [26] was used to integrate Hamilton's equations of motion, supplemented with a Nosé–Hoover thermostatting scheme [24,25], and Gaussian constraints [27,28] to fix the diatomic bond lengths. The integration time step was chosen to be 2.5 fs. Periodic boundary conditions operate in the y-direction. There are no periodic boundary conditions in the x-direction; the ends of the simulation box are dissolving boundaries. That is, if a molecule reaches either of these boundaries it is removed from the simulation.

To ensure that molecular creations are compatible with mass transport, a component of streaming velocity is added to the thermal velocity of newly created molecules. We determine this component of streaming velocity by taking a value for the flux at the center of the flow region and dividing this by the concentration in the appropriate control volume. Since we begin the simulation with zero molecules, initially there will be no measurable flux. As molecules begin to diffuse through the pore, a steady state flux will gradually develop, which is constant at any yz plane in the simulation cell. To prevent the streaming velocity from diverging, we follow Cracknell et al. [12] and introduce a degree of course graining. The method of augmenting the molecular velocities upon creation proceeds as follows: we allow a setting time of 50000 steps for obtaining sufficient molecules in the flow region to obtain a flux. After this time, we divide the flux (averaged over the settling time) by the average concentration in the control volume of interest. This streaming velocity is then added to the thermal component of velocity of newly created molecules in that control volume. After the settling time, the flux is averaged over periods of 1000 molecular dynamics steps and used in the subsequent calculations of the streaming velocity in the control volumes.

Because we employ a smooth potential to represent the graphite planes in a real carbon pore, we need to account for the exchange of momentum that would take place between fluid molecules and the carbon atoms in a real adsorbent. We employ the so-called diffuse boundary condition, based on the diffuse boundary conditions used by Cracknell et al. [12] but modified here for the case of molecules. Application of our molecular diffuse boundary condition proceeds as follows: After each molecular dynamics time step we check to see if the following two conditions are satisfied:

1. The center-of-mass momentum (in the laboratory frame) of a given molecule in the z-direction has reversed in sign;
2. The center-of-mass of that same molecule is within the repulsive region of the Steele 10–4–3 potential.

Table 8.2 Activities used in the DCV GCMD mixture and pure component simulations

| Species | $t = 0(°C)$ | | $t = 25(°C)$ | |
	$Z_{source}(nm^{-3})$	$Z_{sink}(nm^{-3})$	$Z_{source}(nm^{-3})$	$Z_{sink}(nm^{-3})$
Nitrogen	0.3116	0.2083	0.2867	0.1902
Oxygen	0.0780	0.0521	0.0717	0.0479

If, and only if, both of these conditions are satisfied, we reselect the centre-of-mass momentum of that molecule in the directions parallel to the confining surface from a Maxwell–Boltzmann distribution at the appropriate temperature.

One important control parameter in DCV GCMD is the ratio of stochastic to dynamic steps, n_{MC}/n_{MD}. An optimum value for n_{MC}/n_{MD} was arrived at by finding the smallest value, which still yielded the correct concentrations of the two species obtained from conducting separate grand canonical adsorption simulations at the relevant thermodynamic state point for the two control volumes. Using too large a value for this parameter greatly increases the CPU time needed to reach a non-equilibrium steady state. We find that for our simulations a value for n_{MC}/n_{MC} of 50 is optimum.

For the pure component simulations, we used the same activities as for the mixture simulations [4], i.e., the activity gradients of pure oxygen and pure nitrogen were identical to the appropriate activity gradients in the mixture. The set of activities appropriate to the two temperatures studied: $t = 0$ and 25°C are given in Table 8.2. Since the fugacities are the same as the partial fugacities in the mixture, and the bulk gases are close to ideal under these conditions, it follows that the pressure of each pure gas is the same as its partial pressure in the mixture.

A series of simulations was carried out at the following set of pore widths: $H/\Delta = \{5.0, 4.0, 3.0, 2.5, 2.0, 1.9\}$, where Δ is the graphite layer spacing. The length of these simulations ranged from a minimum of 12 million molecular dynamics steps at the highest pore width to about 36 million steps for the lowest pore width.

8.4.3   EMD simulations

Equilibrium molecular dynamics simulations were performed for the purpose of calculating the phenomenological transport coefficients, L_{mn} and L_{f}, plus the coefficient of self-diffusion. The algorithm for the EMD simulations was similar to that used in the dynamic portion of the DCV GCMD simulations with the exception that the overall temperature was thermostatted rather than the local temperature in bins. The diffuse scattering algorithm was also applied in these simulations. A system size consisting of 1000 molecules in total was used throughout. The mean square displacements appearing in Equations (8.4) and (8.6) were calculated along with the molecular square displacements over a 100 ps time frame. Single component EMD simulations were performed at the same set of pore widths and mean densities as used/determined in the DCV GCMD simulations. Mixture simulations were performed at the single pore width of 2.5Δ but at the mean densities and compositions taken from the source and sink control volumes in the mixture DCV GCMD simulations reported in a previous publication [4]. Starting configurations at the appropriate density were obtained by using GCMC to place the molecules inside the pore. These configurations were then equilibrated for 500,000 time steps followed by a production run of 5 million steps, except in the case of the mixture simulations, which were run for 20 million production steps in order to obtain reasonable signal to noise on the cross coefficients.

8.4.4 GCMC simulations

In order to calculate the Darken factors, a series of GCMC simulations were performed in slit-pores having the same range of pore sizes as used in the pure component simulations. In each of these adsorption simulations, the absolute number of molecules within the pore volume was calculated at a range of filling pressures. Each simulation consisted of 30 million attempted Monte Carlo moves (a move can either be a combined rotation-translation, creation or destruction), of which the first 10 million moves were rejected prior to the averaging process. No mixture adsorption simulations were performed.

8.5 Results and discussion

8.5.1 DCV GCMD pure component simulation

Pure component, effective transport diffusivities were obtained from the DCV GCMD simulations by taking the ratio of the total steady state flux and the gradient of the number density. These diffusivities are plotted as a function of pore width and temperature in Figure 8.2. The figure shows that the diffusion coefficient of either species is relatively insensitive to density and temperature at pore widths above 1 nm. The diffusion coefficient of nitrogen is always greater than that of oxygen in this same regime. In the sub-nanometer range of pore widths, diffusivity coefficients become stronger functions of both density and temperature. At a pore width of 0.8375 nm the diffusion coefficients of both oxygen and nitrogen substantially increase, the size of this increase being more marked for the latter species. Lowering the temperature results in a greater increase in diffusivity in both cases. As the pore width is reduced below 0.8375 nm, the diffusion coefficient of nitrogen decreases sharply while that of oxygen decreases a little at 25°C and rises slightly at 0°C before falling again at the lowest pore width. Over a narrow range of pore widths in the sub-nanometer range, the diffusive selectivity changes from a value favouring nitrogen to one favouring oxygen. To understand these differences it is necessary to examine the various contributions to the effective transport diffusivity such as the thermodynamic factors. Equilibrium simulations were performed to enable such an analysis

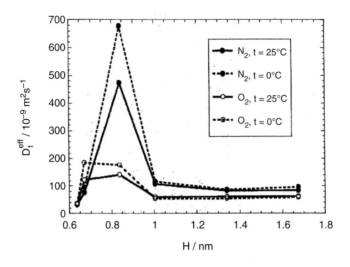

Figure 8.2 Plot of effective transport diffusivity against pore width for nitrogen (solid lines) and oxygen (dashed lines), at $t = 0°C$ (open circles), and $t = 25°C$ (filled circles). Data obtained from pure component DCV GCMD simulations.

to be made. It should be pointed out that the DCV GCMD simulations are still essential for obtaining the total intrapore fluxes since these are needed for the calculation of permeabilities.

8.5.2 EMD pure component simulations

Equilibrium molecular dynamics simulations were conducted for pure oxygen and pure nitrogen as detailed in "Calculation of transport properties via computer simulation." Mean square displacements of the center-of-mass position of the entire fluid and of individual molecules were averaged over the course of these simulations in order to calculate the appropriate diffusion coefficients defined by Equations 8.4 and 8.6.

The self diffusivities are plotted as a function of pore width and temperature in Figure 8.3. From the figure it can be seen that self-diffusivity increases with increasing pore width and temperature for both components. Furthermore, oxygen self-diffusivity is always greater than nitrogen self-diffusivity. These results are in line with the behaviour of D_s in the bulk phase; lowering the pore width corresponds to increasing the density and hence lowering the molecular mobility. Lower temperatures also lead to lower mobility, and hence, lower D_S values. The nitrogen self-diffusivity plots have a small anomaly at $H = 1.005$ nm. Here the self-diffusivity is lower than expected. This can be explained in terms of density. Figure 8.4 shows the density as a function of pore width and temperature. At a pore width of 1.005 nm, the nitrogen density is higher than expected, the effect being stronger at the lower temperature. This feature is absent from the corresponding oxygen

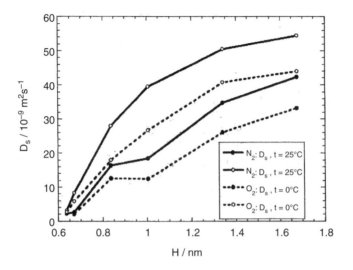

Figure 8.3 Plot of self diffusivity against pore width for nitrogen (filled circles) and oxygen (open circles) at $t = 0°C$ (dashed lines) and $t = 25°C$ (solid lines). Data obtained from pure component EMD simulations.

density plots. The anomalous nitrogen density at $H=1.005$ nm is presumably a molecular packing effect.

The collective diffusion coefficient, D_0, is notoriously difficult to obtain with reasonable signal-to-noise. To illustrate this fact, mean square displacements of the fluid center-of-mass are plotted as a function of pore width for

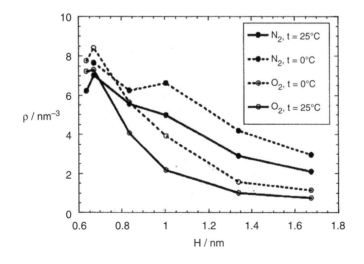

Figure 8.4 Plot of absolute number density against pore width for nitrogen (filled circles) and oxygen (open circles) at $t = 0°C$ (dashed lines) and $t = 25°C$ (solid lines). Densities are those obtained from pure component DCV GCMD simulations.

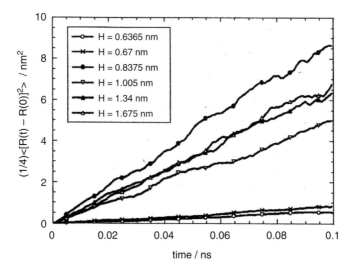

Figure 8.5 Center-of-mass mean square displacements plotted as a function of time for nitrogen at $t = 25°C$ and various pore widths. Data obtained from pure component EMD simulations.

nitrogen at 25°C in Figure 8.5. Except at the narrowest pore widths, the mean square displacement curves are quite noisy despite the use of long simulation times. Figure 8.6 shows a plot of the collective diffusivity. D_0, against pore width for oxygen and nitrogen.

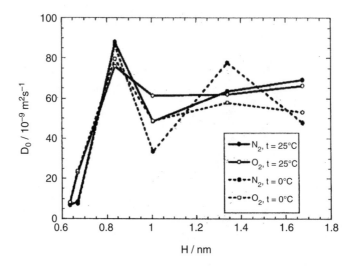

Figure 8.6 Plot of collective diffusivity. D_0, against pore width for nitrogen (filled circles) and oxygen (open circles) at $t = 0°C$ (dashed lines) and $t = 25°C$ (solid lines). Data obtained from pure component EMD simulations.

It is clear that the collective diffusion coefficient has a more complicated dependence on pore width and temperature than the self diffusion coefficient. At 0°C, D_0 is lower for oxygen than at 25°C. This result stands in contrast to the temperature behaviour of the effective transport diffusivity in Figure 8.2. In the case of nitrogen, at some pore widths D_0 is lower at the lower temperature while at others, it is higher at lower temperature. Again, this stands in sharp contrast to the behaviour shown in Figure 8.2. A possible explanation for these differences is the pore width and temperature dependence of the thermodynamic factor (Darken factor). We turn our attention to this in the next section.

Figure 8.6 shows that all the diffusivity curves display a maximum at a pore width of 0.8375 nm in common with the behaviour of the effective transport diffusivity. The collective diffusivity of nitrogen at this point is greater than that of oxygen but this difference is clearly much smaller than seen in the case of D_t^{eff}. Once again we attribute this fact to the thermodynamic factor. At the lower pore width of $H = 0.67$ nm, D_0 for oxygen is greater than that for nitrogen although both diffusivities are substantially lower than they are at 0.8375 nm.

It is of interest to calculate the cross coupling contribution to the collective diffusivity, D_ξ (defined in Equation 8.16. At low loadings, D_ξ is expected to vanish so that the collective and self diffusivities become equal. This fact can be used to calculate transport diffusivities from knowledge of the self-diffusivity and the Darken factor if diffusion occurs at low adsorbate density. However, the densities involved in this work are far removed from the zero loading limit and so the self-diffusivity cannot be expected to equal D_0. Figure 8.7 shows a plot of D_ξ as a function of pore width and temperature for oxygen and nitrogen. The behaviour of D_ξ at 25°C is remarkably similar to the pore width dependence shown by the effective transport diffusivity: insensitivity to pore width above 1 nm, a steep rise at $H = 0.8375$ nm followed

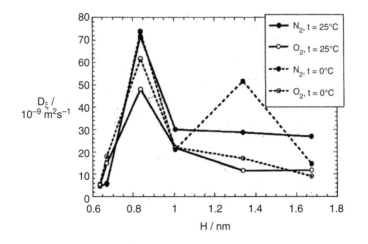

Figure 8.7 Plot of the cross coupling diffusivity, D_ξ, against pore width for nitrogen (filled circles) and oxygen (open circles) at $t = 0$°C (dashed lines) and $t = 25$°C (solid lines). Data obtained from pure component EMD simulations.

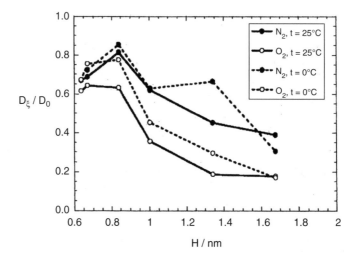

Figure 8.8 Plot of the ratio of the cross coupling diffusivity, D_ξ, to the collective diffusivity, D_0, against pore width for nitrogen (filled circles) and oxygen (open circles) at $t = 0°C$ (dashed lines) and $t = 25°C$ (solid lines). Data obtained from pure component EMD simulations.

by a drop at lower pore widths. The nitrogen D_ξ is greater than the oxygen D_ξ at pore widths greater than 0.67 nm. The pore width behaviour of the effective transport diffusivity results from the pore width dependence of D_ξ. The temperature dependence of D_ξ is complicated. At some pore widths, a lower temperature results in a lower value of D_ξ while at others, it is increased. In the case of nitrogen at 0°C, the diffusivity has two maxima at pore widths of 1.34 and 0.8375 nm. A useful quantity to calculate is the ratio of D_ξ to D_0, which measures the influence of the momentum cross correlations in determining the collective diffusivity. Figure 8.8 shows a plot of this ratio against pore width and temperature for oxygen and nitrogen. The figure shows the increasing importance of this cross correlation diffusivity as the pore width is decreased. Lowering the temperature also increases the contribution from D_ξ relative to the self diffusivity. Comparing Figures 8.4 and 8.8, we see that increasing adsorbate density is chiefly responsible for the growing contribution made by D_ξ to D_0. It is clear from Figure 8.8 that the product of the self diffusivity and the Darken factor would seriously underestimate the value of D_0 and, as a consequence, miss the important pore width behaviour of the transport diffusivity. The increase in D_ξ/D_0 with decreasing temperature reflects the greater adsorbate density at the lower temperature.

In all the pure component data presented so far, the density varied along with pore width. In order to look at the effects of these two variables separately, we conducted EMD simulations at the lowest two pore widths, with $t = 25°C$, for a range of densities. The results of these simulations are shown in Figures 8.9 and 8.10. In Figure 8.9 we see that D_0 for nitrogen is greater than D_0 for oxygen across the entire density range. The oxygen D_0 value is

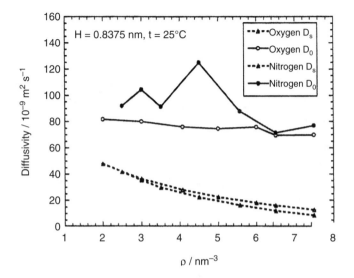

Figure 8.9 Plot of self diffusivity, D_s (dashed lines), and collective diffusivity, D_0 (solid lines), against density at a pore width of $H = 0.8375$ nm ($= 2.5\Delta$) and temperature $t = 25°C$. The different symbols correspond to nitrogen (filled symbols) and oxygen (open symbols). Data obtained from pure component EMD simulations.

only weakly density dependent whereas that for nitrogen shows two maxima occuring at densities of 3 and 4.5 nm^{-3}. The latter density gives rise to a maximum nitrogen diffusive selectivity. The self-diffusivities show a small selectivity toward oxygen at all densities studied. Turning now to Figure 8.10, the D_0 selectivity is now inverted so that oxygen diffuses faster than nitrogen at all densities. Furthermore, the degree of selectivity is much greater than it was at the wider pore width. The oxygen D_0 value increases with decreasing density (with the exception of a small increase at the highest density) and does not go through a maximum, while the nitrogen D_0 value is virtually independent of density. At 3 nm^{-3}, the D_0 selectivity is about 4 while the D_S selectivity is around 5. The contrast between the diffusivities and their density dependence in Figures 8.9 and 8.10 suggests different diffusion mechanisms are in operation at the two different pore widths.

8.5.3 GCMC adsorption simulations

The difference between the transport diffusivity and the collective diffusivity is a thermodynamic multiplication factor known as the Darken factor (Equation 8.15). Much of the density dependence of D_t is associated with this factor. It is therefore important to know what that density dependence is as a function of pore width. Adsorption isotherms were generated for both pure component fluids at the two temperatures of interest by GCMC. A selection of these isotherms at 25°C is plotted in Figure 8.11a. All isotherms

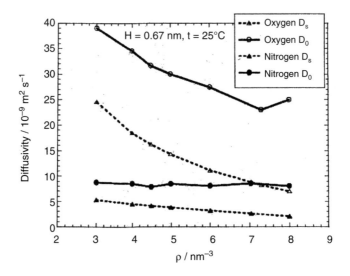

Figure 8.10 Plot of self diffusivity, D_s (dashed lines), and collective diffusivity. D_0 (solid lines), against density at a pore width of $H = 0.67$ nm ($= 2\Delta$) and temperature $t = 25°C$. The different symbols correspond to nitrogen (filled symbols) and oxygen (open symbols). Data obtained from pure component EMD simulations.

are simple type I isotherms. Outside of the Henry law region, at a given filling pressure, we see that oxygen is more strongly adsorbed than nitrogen. As the pore width is reduced, the amount of either species of gas adsorbed increases at the lower pressures. This latter observation is a direct result of the increased overlap of the potential energy surfaces of both graphite planes as they move closer together. Greater overlap results in deeper potential energy wells, which leads to greater adsorption. At high filling pressures, beyond monolayer coverage, entropic effects dominate the adsorption process. The smaller oxygen molecule is more easily accommodated than the slightly bulkier nitrogen molecule. At the low filling pressure used in the DCV GCMD simulations (see Figure 8.11b), the differences between nitrogen and oxygen adsorption are less significant. Indeed, apart from the lowest pore widths, nitrogen is more strongly adsorbed than oxygen. At the lowest pore width, entropic effects once again dominate which favours the smaller oxygen molecules. Based on these results, we can speculate that adsorption selectivity in a mixture of the two gases would be small at the operating pressures used in DCV GCMD simulations but would increase in favour of oxygen at high pressures and very low pore widths.

In order to calculate the Darken factors, we fitted our isotherm data to the following equation

$$\ln f = A + \ln \rho + B_1\rho + B_3\rho^3 + B_5\rho^5 \tag{8.22}$$

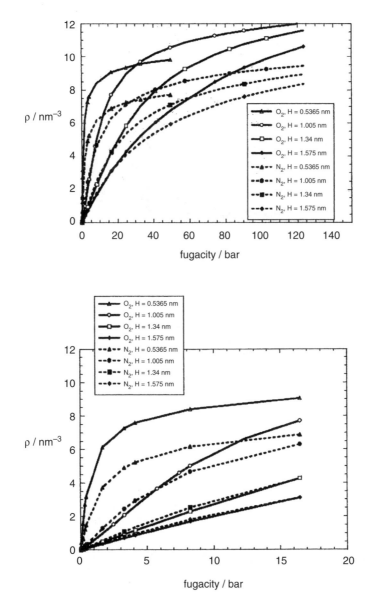

Figure 8.11 (a) Adsorption isotherms plotted at various pore widths for oxygen (solid lines and open symbols) and nitrogen (broken lines and filled symbols) at a temperature, $t = 25°C$. The symbols used in the figure correspond to: $H = 0.6365$ nm (triangles), $H = 1.005$ nm (circles), $H = 1.34$ nm (squares) and $H = 1.675$ nm (diamonds). Data obtained from equilibrium GCMC simulations. (b) As for (a), but showing the 0–20 bar fugacity regime in more detail.

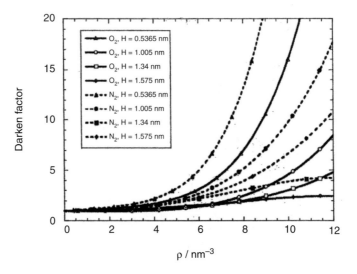

Figure 8.12 Plot of Darken factors (dln f/ dln ρ), against density for oxygen (solid lines and open symbols) and nitrogen (broken lines and filled symbols), at a temperature, $t = 25°C$. The symbols used in the figure correspond to: $H = 0.6365$ nm (triangles), $H = 1.005$ nm (circles), $H = 1.34$ nm (squares) and $H = 1.675$ nm (diamonds). These curves were generated using Equation (8.23).

where B_1, B_3 and B_5 are empirical parameters. The expression for the Darken factor follows from differentiating Equation (8.22) with respect to $\ln \rho$. This gives

$$\frac{\text{dln} f}{\text{dln} \rho} = 1 + B_1\rho + 3B_3\rho^3 + 5B_5\rho^5 \qquad (8.23)$$

from which it can be seen that the limiting value of the Darken factor as $p(\rightarrow 0$ is unity. Darken factors have been generated for a range of densities at each pore width based upon the values of the coefficients B_1–B_5 obtained in the above fitting procedure. A selection of these factors at 25°C is plotted in Figure 8.12 as a function of density. From the figure we see that at any given density, the Darken factor increases as pore width decreases with the exception of the nitrogen factor between pore widths of 1.34 and 1.675 nm where this trend is reversed. At a given pore width, the Darken factor for nitrogen is greater than that of oxygen across a wide range of density. There is one exception to this trend. At the pore width of 1.34 nm, the Darken curves cross at a density of 11.5 nm^{-3} such that oxygen has the greater Darken factor at higher densities. We cannot attach too much significance to this anomalous behaviour since the Darken factors have been extrapolated to densities higher than those found in the GCMC simulations.

In Figure 8.13, various dynamic selectivity measures are plotted against pore width. These selectivity measures are defined as the ratio of an oxygen diffusivity and nitrogen diffusivity or oxygen Darken factor and nitrogen

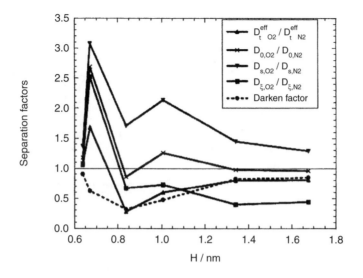

Figure 8.13 Plot of various separation factors (defined as the ratio of an oxygen diffusivity or Darken factor to the nitrogen diffusivity or Darken factor) against pore width at $t = 25°C$.

Darken factor. Taking first the collective diffusivity ratio, $D_{0,O_2}/D_{0,N_2}$, we see that this factor indicates a maximum selectivity for nitrogen at a pore width of 0.8375 nm and a maximum selectivity for oxygen at the lower pore width of 0.67 nm. The Darken factor ratio works in favour of nitrogen at all pore widths although it is a maximum for nitrogen at 0.8375 nm. It is the Darken factor contribution, which strongly enhances the dynamic selectivity at this pore width, as can be seen by comparing the collective diffusivity ratio, $D_{0,O_2}/D_{0,N_2}$, with the effective transport diffusivity ratio, $D^{eff}_{t,O_2}/D^{eff}_{t,N_2}$. The Darken factor works against oxygen at 0.67 nm, lowering the oxygen selectivity from 2.7 to 1.7 in the case of the effective transport diffusivity.

We now address the viscous contribution to the total flux. As detailed elsewhere, the diffusion coefficients obtained from our DCV GCMD simulations are strictly speaking, only *effective* transport diffusivities. To obtain the true transport diffusivity, one can either use the viscous subtraction method [19] to yield the diffusive flux by NEMD, or one can multiply the collective diffusivity by the Darken factor (equilibrium route). The difference between the DCV GCMD effective transport diffusivity and the equilibrium calculated transport diffusivity is a measure of the viscous contribution to the flow. Figure 8.14 shows a plot of both transport diffusivities as a function of pore width at 25°C. As the figure shows, the difference between the two diffusivities is insignificant. At one or two pore widths, the transport diffusivity of oxygen appears to be greater than the effective transport diffusivity, which is counterintuitive. We believe this simply reflects the statistical uncertainties in the former diffusivity relative to the latter. For pores in the range we have studied it is safe to assume that the viscous contribution to flow is weak in comparison

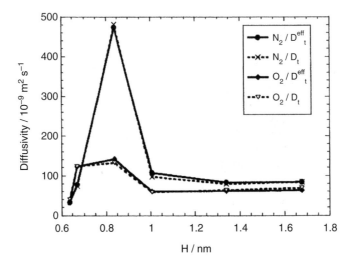

Figure 8.14   Comparison of the effective transport diffusivity (solid lines) obtained via DCV GCMD and the transport diffusion coefficient obtained from the product of the collective diffusivity and the Darken factor (broken lines). Data shown refer to a temperature of $t = 25°C$.

to the diffusive contribution. To a good approximation, the effective transport diffusivity may be used instead of the true transport diffusivity.

8.5.4 *Comparison of mixture and pure component data*

In a previous publication [4] we reported DCV GCMD results for an 80:20 mixture of nitrogen and oxygen diffusing through graphite slits identical to those in our current study. In order to compare diffusion coefficients of either species in the mixture with those of the pure components, we have performed EMD simulations with such an 80:20 mixture at 25°C and a pore width of 0.8375 nm in order to calculate the Stefan–Maxwell coefficients in the mixture. Recall that the DCV GCMD mixture simulations cannot yield these coefficients unambiguously. Because the composition and density in the source and sink control volumes differed in the DCV GCMD mixture simulations [4], separate EMD simulations were performed at these compositions and densities to see what effect this had on the results. The density and composition in the source was: $\rho = 6.044$ nm^{-3}, $x_{N2} = 0.18$, while in the sink it was: $\rho = 5.644$ nm^{-3}, $x_{N2} = 0.17$. These simulations had to be run for a total of 20 million time steps each so that we had a good signal to noise ratio on the data. Because of the extremely long simulation times, we conducted these runs at a single pore width.

The mean square displacements plotted against time from the higher density simulation are shown in Figure 8.15. This figure gives an idea of the accuracy of our data. The mean square displacements are reasonably linear in

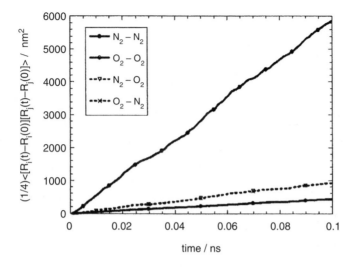

Figure 8.15 Plot of mean square displacements versus time for the 80:20 mixture at $t = 25°C$ and $H = 0.8375$ nm. Data obtained from EMD simulations.

time. The nitrogen–nitrogen curve shows a much steeper gradient than either the oxygen–oxygen, or the cross coupling curves. The phenomenological coefficients L_{ij} obtained from the slopes of these curves are given in Table 8.3. Here $1 = O_2, 2 = N_2$. The first thing to note about these quantities is that within the statistical uncertainties, $L_{12} = L_{21}$, in agreement with Onsager's regression hypothesis. Second, L_{11} is an order of magnitude greater than L_{22} while the cross coefficient L_{12} is a little over twice as large as L_{22}. Clearly, the cross diffusion coefficient is significant in magnitude. The L_{ij} coefficients were converted into Stefan-Maxwell mutual diffusion coefficients using Equations 8.8 and 8.9, in which L_x is taken to be the mean value of L_{12} and L_{21}, the results being collected in Table 8.4. From Table 8.4 we see that D_{1M} is almost twice the magnitude as D_{2M} at the higher density, but only 25% larger at the lower, sink density. In order to compare D_{1M} and D_{2M} with the D_0 values obtained

Table 8.3 Phenomenological transport coefficients for the 80:20 mixture of nitrogen and oxygen at t = 25°C obtained using equation 8.4.

	Source	Sink
$k_B V L_{11}$ (10⁻⁶m²s⁻¹)	59.370(93)	55.83(14)
$k_B V L_{12}$ (10⁻⁶m²s⁻¹)	9.342(32)	7.994(23)
$k_B V L_{21}$ (10⁻⁶m²s⁻¹)	9.333(36)	7.982(22)
$k_B V L_{22}$ (10⁻⁶m²s⁻¹)	4.234(7)	4.371(5)

Note: The quantities in parentheses are the statistical uncertainties in the last digits and represent the error in the slope of the linear portion of the mean square displacement plots. The subscripts 1 and 2 refer to nitrogen and oxygen, respectively

Table 8.4 Stefan-Maxwell coefficients for the 80:20 mixture of nitrogen and oxygen at $t = 25°C$ obtained using equations 8.8 and 8.9 with L_{12} and L_{21} symmetrized.

	Source	Sink
D_{1M} (10^{-9}m²s⁻¹)	91.667	79.392
D_{2M} (10^{-9}m²s⁻¹)	54.184	63.001
D_x (10^{-9}m²s⁻¹)	17.581	22.564

Note: The subscripts 1 and 2 refer to nitrogen and oxygen, respectively. The quantities in parentheses are the statistical uncertainties in the last digits

from the pure component simulations at the same temperature and pore width, we first average the values for the source and sink conditions. For the mixture we have $\bar{D}_{1M} = 86 \times 10^{-9}$ m² s⁻¹, $\bar{D}_{2M} = 59 \times 10^{-9}$ m² s⁻¹, while for the pure components we have $D_0^{N_2} = 88 \times 10^{-9}$ m² s⁻¹, $D_0^{O_2} = 76 \times 10^{-9}$ m² s⁻¹. From this comparison we note that in the case of nitrogen, its diffusion through the slit pore is only marginally effected by the presence of the oxygen component. Oxygen diffusion on the other hand is significantly reduced in the mixture at this pore width and temperature. One important observation from this set of simulations is that L_x (and hence D_x), the cross diffusion coefficient, is non-negligible. This will obviously have important consequences for the total intrapore fluxes. The nitrogen flux is essentially determined by the L_{11}, coefficient while the oxygen flux is determined by contributions from both L_{22} and L_x. We note that MacElroy and Boyle observed that the diffusion cross coupling for methane–hydrogen mixtures was weak [29].

The performance of a membrane in separating gas mixtures is frequently discussed in terms of the permeability of a given gas species. The permeability, F, which is the pressure and thickness normalized flux, is defined by

$$F_i = \frac{J_i}{(\Delta p_i / L)} \tag{8.24}$$

where, J_i is the molecular flux of component i and Δp is the pressure drop across a membrane of length L. In the mixture simulations, the partial pressure drops of nitrogen and oxygen were 4 bar and 1 bar, respectively. The pure component pressure drops were chosen to be the same as these mixture partial pressure drops.

A selectivity measure based on the species permeabilities, the so-called permselectivity, is defined by

$$\alpha_{O_2/N_2} = \frac{F_{O_2}}{F_{N_2}} \tag{8.25}$$

Permselectivity values are plotted in Figure 8.16 for both pure component data and mixture data as a function of pore width. The main feature

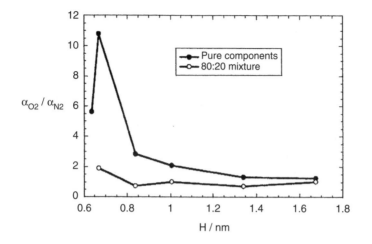

Figure 8.16 Plot of permselectivity (α_{O_2}/N_2) against pore width at $t = 25°C$. The filled symbols represent the permselectivity for the pure components while the open symbols represent the mixture permselectivity. Data obtained from DCV GCMD simulations (mixture results are taken from Ref. [5]).

of Figure 8.16 is the large difference between the pure component and mixture selectivities. In the mixture case, oxygen selectivity varies weakly with pore width, reaching a value of about 2 at the lowest pore width studied. In the case of the pure components, oxygen selectivity is higher than in the mixture at all pore widths. The oxygen selectivity reaches a maximum value of about 11 at a pore width of 0.67 nm, exactly the same pore width at which the diffusive selectivity is a maximum for oxygen. At the lowest pore width studied, oxygen selectivity decreases, again mirroring the behaviour of the diffusive selectivity.

8.5.5 *Possible diffusion mechanisms*

Our results for the pure component diffusion have revealed a selectivity reversal at low pore widths. Clearly two different mechanisms give rise to these different regimes. The pore width below which oxygen becomes selective is 0.8375 nm. This pore width is wide enough (allowing for the dead space due to the carbon atoms in graphite) for both molecules to rotate about both axes. The length of a nitrogen molecule is 0.439 nm while that of an oxygen molecule is 0.411 nm. At the next lowest pore width (of physical width 0.67 nm), the nitrogen molecules would be unable to rotate about one of their axes. The oxygen molecules, while they could not rotate freely about the same axis, may still undergo large amplitude "frustrated" rotations. Using transition state theory, Singh and Koros [30] have shown that such a loss of rotational freedom in nitrogen can indeed lead to a drop in diffusivity relative to that of oxygen. We therefore postulate that such "entropic" effects

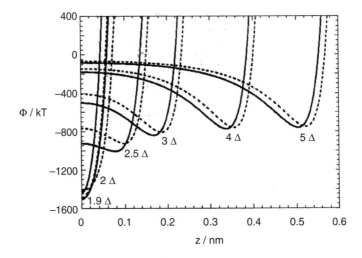

Figure 8.17 Plot of graphite-molecule intermolecular potential energy as a function of the z-coordinate (as the pore is symmetric we show only the positive values of z). The energy is plotted at six different pore widths for nitrogen (solid lines) and oxygen (broken lines).

are responsible for the selectivity observed at 0.67 nm. Below this pore width, oxygen will loose its "entropic" advantage at which point selectivity is based upon molecular size.

At 0.8375 nm, both diffusion coefficients are maximised but selectivity is now in favour of nitrogen. In order to understand this phenomenon, it is necessary to look at the behaviour of the intermolecular energy between a molecule and the graphite planes as a function of pore width. Average potential energies were obtained in a simulation by taking a single molecule and randomly translating and rotating it within the pore space such that the potential energy was averaged over all positions and orientations. Figure 8.17 shows the potential energy plotted as a function of the z-co-ordinate for the entire range of pore widths studied. What we notice from this figure is a substantial lowering of the potential barrier height for nitrogen at 0.8375 nm, compared to the higher pore widths. Figure 8.18 shows the 0.8375 nm pore width potentials in more detail. If molecules are to hop from one graphite plane to the opposing plane, they must have sufficient thermal energy to overcome this potential barrier. At pore widths lower than 0.8375 nm, molecules are trapped in a deep potential well between the two walls from which they cannot escape. Conditions at 0.8375 nm are optimum, however, for the wall to wall hopping mechanism to occur. Beyond this pore width, the barrier heights are too great and the majority of the molecules spend most of their time trapped in the vicinity of one wall or another. This regime is then characterised by an energetic selectivity mechanism as opposed to the entropic selectivity mechanism in operation in pores less than or equal to 0.67 nm in width.

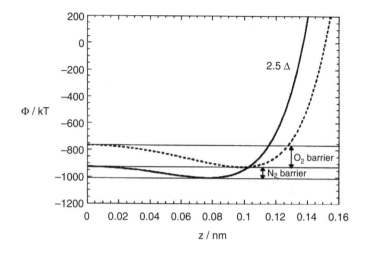

Figure 8.18 Plot of graphite-molecule intermolecular potential energy as a function of the z-coordinate (as the pore is symmetric we show only the positive values of z). The energy is plotted at a pore width of 0.8375 nm to highlight the difference between the nitrogen (solid lines) and oxygen (broken lines) energies.

8.6 Summary and conclusions

Using the industrial separation of air into its major constituents as a motivation we have undertaken a molecular simulation study of the mass transport of oxygen and nitrogen through graphite slit pores in a bid to understand the molecular origins of the reported kinetic oxygen selectivity. We have used both non-equilibrium molecular dynamics methods and equilibrium methods (Monte Carlo and molecular dynamics) in our study to extract the maximum information. The method of DCV GCMD has the advantage of enabling a direct calculation of permeability and permselectivity in a simulation that closely mimics gas flow though membranes under pressure and chemical potential gradients. The disadvantage of this method is that the diffusion coefficients are not easily obtained, particularly in the case of mixture diffusion. These limitations are offset by using the EMD and GCMC methods.

Our results show that a permselectivity in favour of oxygen is obtained across a range of pore widths for pure gas transport. This selectivity is significantly reduced when oxygen is present as the minor constituent in a mixture of oxygen and nitrogen at a composition similar to that in air. Furthermore, the permselectivity inverts at some pore widths to favour nitrogen. The permselectivity can be split into two contributions: a diffusive selectivity and a sorptivity selectivity. The latter quantity is a thermodynamic quantity that is related to the inverse slopes of the adsorption isotherms. Using GCMC we have explored the pore width and density dependence of the related quantity, the Darken factor, and found that the ratio of Darken factors is less than unity. Thus, the sorptivity favours nitrogen at all pore widths and densities studied, although the ratio becomes closer to unity at lower pore widths and may exceed it at yet lower pore widths. The other contribution to the permselectivity arises from the diffusivity ratio.

Using EMD we have been able to establish that viscous contributions to the intrapore flux are weak. These simulations established that the collective diffusivities, D_0, show an interesting pore width dependence. There are three diffusive regimes: A regime at pore widths in excess of 1 nm in which the diffusion coefficients show a weak pore width dependence at ambient temperature, a second regime at around 0.8375 nm where the diffusion coefficients increase, and a third regime at narrower pore widths where the diffusion coefficients decrease sharply. This latter regime is characterised by a diffusive selectivity in favour of oxygen, which we believe to be an "entropic" effect in agreement with the findings of Singh and Koros [30]. This "entropic" effect arises in pores which are too narrow for the larger nitrogen molecules to rotate freely about both axes but not narrow enough to prevent the oxygen molecules from rotating about both axes. The sharp drop in the absolute values of the diffusivities below 1 nm corresponds to the adsorbate becoming a quasi-two-dimensional fluid; the slits are too narrow for more than one layer of molecules to form in the z-direction. These highly confined molecules are trapped in deep potential energy wells from which they cannot execute hopping motions. The lack of sensitivity of the diffusivities to pore width for the wider pores is a result of the molecules being tightly bound to the graphite surfaces. The barriers over which the molecules must hop to reach the alternative surface are too great at ambient temperature. These barrier heights do not significantly reduce until a pore width of 0.8375 nm is reached, at which point the barrier for nitrogen is significantly lower than that of oxygen. This fact may explain why there is nitrogen selectivity at this pore width. At pore widths lower than 0.67 nm, the extra rotational degree of freedom possessed by the oxygen molecule is lost. Size selectivity then becomes important. The smaller oxygen molecules will diffuse faster than nitrogen but the absolute value will be low. At the lowest pore width we have studied, there is evidence of such an effect—oxygen is adsorbed more strongly than nitrogen at low pressures as a result of the nitrogen–graphite potential energy being shifted upwards as the repulsive energy begins to dominate. This marks the beginning of the molecular sieving regime.

The use of EMD simulations has allowed us to split the collective diffusivity into a self diffusivity, D_S, and a cross coupling diffusivity, D_ξ. We find that the self-diffusion contribution to the collective diffusion coefficient decreases as the pore width decreases. As pore width decreases, density increases. The trend in D_S with density follows closely the trend seen in bulk fluids. The cross coupling diffusivity, on the other hand, varies with pore width, and begins to dominate the self diffusivity as the pore width narrows. This effect is not simply related to the density change. The cross coupling diffusivity is greater for nitrogen at pore widths larger than 0.67 nm and this cancels out any self diffusion selectivity in favour of oxygen. At lower pressures, where transport diffusion is dominated by self diffusion, higher oxygen selectivities will result while at higher pressures, dominated by cross coupling diffusivity, nitrogen selectivity will result unless the pore widths become so narrow that entropic effects work against nitrogen.

Our simulation study has revealed a rich behaviour of diffusion coefficients for oxygen and nitrogen in single slit pores. Real molecular sieving carbon (MSC) contains a distribution of pore widths. Typical MSC pore size distributions display a maximum around 0.5 nm. Allowing for the excluded volume of the implicit carbon atoms in our slit pore model, the pores of width 0.8375 nm and lower correspond to typical pores found in MSC. In this regime of pore widths, we observe a large change in diffusion behaviour. High diffusivities at 0.8375 nm mean high fluxes, albeit at the expense of oxygen selectivity. Lower pore widths favour oxygen selectivity but at the expense of greatly lowered absolute diffusivity and hence lower permeability. However, by lowering the pressure, both the absolute oxygen diffusivity and the oxygen selectivity can be improved since pressure has virtually no effect on the nitrogen diffusivity at the 0.67 nm pore width.

High oxygen selectivity for pure component flow does not necessarily translate to high selectivity in mixtures of gases. We have shown that the oxygen selectivity is severely reduced in the 80:20 mixtures. Furthermore, we have clearly demonstrated that the cross diffusion coefficients in mixtures are not insignificant and must therefore be taken into consideration in any model of mass transfer through membranes. A knowledge of the composition and pore width dependence of these cross diffusion coefficients and how they relate to the pure component diffusion coefficients is the subject of our future work.

Finally, we note that our simple model is unable to reproduce the large oxygen selectivities obtained in experimental studies of uptake in real molecular sieving carbons. The single slit pore model, whilst a very convenient and useful theoretical construction, is probably too crude to capture the real effects of a microporous carbon, where many adsorbate molecules will be in connections or at edges of microcrystals. Simulation results of gas transport through pores based on the randomly etched graphite pore model (REGP) of Seaton et al.[2] suggest that attention should be focused towards studying transport of oxygen and nitrogen in more realistic microporous carbon models such as those generated via reverse Monte Carlo [31].

Acknowledgments

We are grateful to the National Science Foundation for the support of this work through research grant no. CTS-9908535; supercomputer time was provided through an NSF NRAC grant no. MCA93SOIIP.

References

1. Chihara, K. and Suzuki, M. (1979) "Control of micropore diffusivities of molecular sieving carbon by deposition of hydrocarbons," *Carbon* 17, 339.
2. Seaton, N.A., Friedman, S.P., MacElroy, J.M.D. and Murphy, B.J. (1997) "The molecular sieving mechanism in carbon molecular sieves: a molecular dynamics and critical path analysis," *Langmuir* 13, 1199.
3. MacElroy, J.M.D., Friedman, S.P. and Seaton, N.A. (1999) "On the origin of transport resistances within carbon molecular sieves," *Chem. Engng Sci.* 54, 1015.
4. Travis, K.P. and Gubbins, K.E. (1999) "Transport diffusion of oxygen–nitrogen mixtures in graphite pores: a nonequilibrium molecular dynamics (NEMD) study," *Langmuir* 15, 6050.
5. Mason, E.A. and Viehland, L.A. (1978) "Statistical–mechanical theory of membrane transport for multicomponent systems: passive transport through open membranes," *J. Chem. Phys.* 68, 3562.
6. Mason, E.A. and Malinauskas, A.P. (1983) *Gas Transport in Porous Media: The Dusty Gas Model* (Elsevier, Amsterdam).
7. MacElroy, J.M.D. (1996) "Diffusion in homogeneous media," in *Diffusion in Polymers*, Neogi, P., Ed, (Marcel Dekker, New York).
8. Karger, J. and Ruthven, D.M. (1992) *Diffusion in Zeolites and Other Microporous Solids* (Wiley, New York).
9. Nicholson, D. (1998) "Simulation studies of methane transport in model graphite micropores," *Carbon* 36, 1511.
10. Heffelfinger, G. and Swol, F.V. (1994) "Diffusion in Lennard–Jones fluids using dual control volume grand canonical molecular dynamics simulation (DCV-GCMD)," *J. Chem. Phys.* 100, 7548.
11. MacElroy, J.M.D. (1994) "Nonequilibrium molecular dynamics simulation of diffusion and flow in thin microporous membranes," *J. Chem. Phys.* 101, 5274.
12. Cracknell, R.F., Nicholson, D. and Quirke, N. (1995) "Direct molecular dynamics simulation of flow down a chemical potential gradient in a slit-shaped micropore," *Phys. Rev. Lett.* 74, 2463.
13. Pohl, P.I., Heffelfinger, G.S. and Smith, D.M. (1996) "Molecular dynamics computer simulation of gas permeation in thin silicalite membranes," *Mol. Phys.* 89, 1725.
14. Ford, D.M. and Heffelfinger, G.S. (1998) "Massively parallel dual control volume grand canonical molecular dynamics with LADERA II. Gradient driven diffusion through polymers," *Mol. Phys.* 94, 673.
15. Travis, K.P. and Gubbins, K.E. (1998) *Sixth Fundamentals of Adsorption* (Elsevier, Amsterdam), pp 1161–1166.
16. Nicholson, D. (1997) "The transport of adsorbate mixtures in porous materials: basic equations for pores with simple geometry," *J. Membr. Sci.* 129, 209.
17. Travis, K.P., Todd, B.D. and Evans, D.J. (1997) "Departure from Navier–Stokes, hydrodynamics in confined liquids," *Phys. Rev. E* 55, 4288.
18. Travis, K.P. and Gubbins, K.E. (2000) "Poiseuille flow of Lennard–Jones fluids in narrow slit pores," *J. Chem. Phys.* 112, 1984.
19. Travis, K.P. and Gubbins, K.E. (2000) "Combined diffusive and viscous transport of methane in a carbon slit pore," *Mol. Simul.* 25, 209.
20. Steele, W.A. (1974) *The Interaction of Gases with Solid Surfaces* (Pergamon Press, Oxford).

21. Weiner, S.J., Kollman, P.A., Nguyen, D.T. and Case, D.A. (1986) "An all atom force field for simulation of proteins and nucleic acids," *J. Comput. Chem.* 7, 230.
22. Beenakker, J.J.M., Bonnan, V.D. and Krylov, S.Y. (1995) "Molecular transport in subnanometer pores: zero-point energy, reduced dimensionality and quantum sieving," *Chem. Phys. Letts.* 232, 379.
23. Kaneko, K., Cracknell, R.F. and Nicholson, D. (1994) "Nitrogen adsorption in slit pores at ambient temperatures: comparison of simulation and experiment," *Langmuir* 10, 4606.
24. Nose, S. (1984) "A unified formulation of the constant temperature molecular dynamics method," *J. Chem. Phys.* 81, 511.
25. Hoover, W.G. (1985) "Canonical dynamics—equilibrium phase-space distributions," *Phys. Rev. A* 31, 1695.
26. Alien, M.P. and Tildesley, D.J. (1987) *Computer Simulation of Liquids* (Oxford Science Publications, Oxford).
27. Evans, D.J. and Morriss, G.P. (1990) *Statistical Mechanics of Nonequilibrium Liquids* (Academic Press, London).
28. Daivis, P.J., Evans, D.J. and Morriss, G.P. (1992) "Computer simulation study of the comparative rheology of branched and linear alkanes," *J. Chem. Phys.* 97, 616.
29. MacElroy, J.M.D. and Boyle, M.J. (1999) "Nonequilibrium molecular dynamics simulation of a model carbon membrane separation of CH_4/H_2 mixtures," *Chem. Engng. J.* 74, 85.
30. Singh, A. and Koros, W. (1996) "Significance of entropic selectivity for advanced gas separation membranes," *J. Ind. Engng. Chem. Res.* 35, 1231.
31. Tomson, K.T. and Gubbins, K.E. (2000) "Modeling structural morphology of microporous carbons by reverse Monte Carlo," *Langmuir* 16, 5761.

chapter nine

Simulation study of sorption of CO_2 and N_2 with application to the characterization of carbon adsorbents

S. Samios

G.K. Papadopoulos*

T. Steriotis

A.K. Stubos

National Center for Scientific Research Demokritos

Contents

* Corresponding author.
Reprint from *Molecular Simulation*, 27: 5–6, 2001. http://www.tandf.co.uk

9.1   Introduction

Pore structure characterization is an important prerequisite for the selection and efficient utilization of porous adsorbents and catalysts in a number of industrial applications including separation processes, removal of various pollutants and gas storage. In the case of mesoporous and macroporous materials, there exist several more or less established characterization methods that provide information on pore size distribution (PSD), pore network connectivity and other structural parameters of the material [1,2]. On the contrary, the reliable assessment of microporosity (pores of sizes less than 2 nm) in terms of relating sorption properties to the underlying microstructure is much less advanced. The commonly used Dubinin–Radushkevich, Dubinin–Astakhov and Dubinin–Stoeckli methods employ phenomenological models of adsorption based on the thermodynamic approach of Dubinin. The limitations of these and other conventional methods used in practice (like the MP and the Horvath–Kawazoe methods) for micropore size characterization have been repeatedly discussed in the literature (see for example [4,5] and related references therein, see also [6] for a recent assessment of different techniques for the estimation of PSD in carbons). The criticisms raised are related mainly to the fact that the mechanism of molecular adsorption in micropores is still under active debate.

Improved approaches to the micropore structure characterization problem have been recently developed based on molecular level theories and statistical mechanics based simulations. In particular, density functional theory (DFT) in a sufficiently elaborate form has been used to provide an accurate description of simple fluids in geometrically simple confined spaces and develop practical methods for the evaluation of the pore structure over a wide range of pore sizes [7–12]. To capture more accurately the behavior of the adsorbates in micropores, it is often necessary to model them as non-spherical molecules with electrostatic interactions. Given the limited capabilities of DFT in this context, molecular simulation based on the Grand Canonical Monte Carlo technique has been established lately as an efficient alternative approach for the generation of adsorption isotherms in carbons and the subsequent determination of PSDs [13–21]. Some authors have combined these studies with structural investigations for the densification process in carbon nanopores using spherical molecules, ethane and carbon dioxide and accounting for effects of pore shape and size, temperature, quadrupole interactions and molecule length [19,22–24].

Use of GCMC method for obtaining the PSD of microporous carbonaceous materials involves the following three major steps [13,20,21]:

1. Determination (and validation whenever possible) of a molecular model for the adsorbate–adsorbate and adsorbate–adsorbent interactions.
2. Generation of a database of sorption isotherms with respect to a specific adsorbate for a set of pore widths, pressures and temperatures.

3. Inversion of the adsorption integral equation:

$$N(p) = \int_{H_{\min}}^{H_{\max}} f(H)n(H)\mathrm{d}H \tag{9.1}$$

where $N(p)$ is the experimentally measured amount of adsorbate, $n(H,P)$ is the average density of adsorbate at pressure p in a pore of width H, and $f(H)$ is the PSD sought.

The solution of Equation 9.1 is an ill-posed problem. Depending on the form of the kernel $n(H,p)$ and the isotherm $N(p)$, there can be from zero to an infinity of solutions for $f(w)$ (detailed discussions on the methods for the solution of Equation 9.1 and the application of suitable constraints to force physically sound or appealing solutions including constraints on the smoothness of $f(H)$ and the range of H, see [14,18,20,21] and references therein). Nonetheless, our work aims at finding useful solutions to Equation 9.1 in the sense that the gas adsorption properties of microporous carbons can be reliably predicted. For instance, our efforts concentrate on predicting gas adsorption isotherms for various adsorbents and temperatures from a PSD obtained with a particular gas at a given temperature.

Neimark and co-workers have also used N$_2$ and Ar at 77 K and CO$_2$ at 273 K as adsorbates and generated $n(H,p)$, based on DFT methods (for N$_2$ and Ar) and GCMC for CO$_2$ up to the pressure of 1 bar employing the Harris and Yung model [25] for adsorbate–adsorbate interactions. They found reasonable agreement between PSDs determined with the different gases on various porous carbon samples. In addition, they reported satisfactory comparisons between PSDs of microporous carbons determined from the DFT and GCMC databases for CO$_2$.

Sweatman and Quirke [21] have employed molecular simulation techniques including the Gibbs method Monte Carlo to determine the molecular models for N$_2$, CH$_4$ and CO$_2$, as well as GCMC simulations to generate adsorption isotherms in carbon slit-pores at 298 K for pressures up to 20 bar. These data have then been used for the calculation of PSDs for typical activated carbons. They found that the high pressure measurements of CO$_2$ reveal micropore structure not seen with the other gases or with measurements up to 1 bar. Their results also indicate that the CO$_2$ based PSDs are the most robust in the sense that they can predict the adsorption of methane and nitrogen at the same ambient temperature with reasonable accuracy.

The work presented in this chapter builds upon previous work by Samios et al. [13,19] who used GCMC in combination with CO$_2$ experimental isotherm data at 195.5 K and ambient temperatures to characterize microporous carbons and obtain the corresponding PSDs. Specifically, the databases $n(H,p)$ have been built by determining the mean CO$_2$ density inside single slit-shaped graphitic pores of given width (from 0.5 to 2.0 nm) along with utilization of N$_2$ at 77 K. High pressure data for CO$_2$ are used as well and the isosteric heat of adsorption is employed to further validate the obtained PSDs.

9.2   Modeling of the molecular interactions

9.2.1   Adsorbates

Carbon dioxide is modeled as a three charged center molecule, according to Murthy and co-workers [26] with the parameters $\varepsilon_{OO}/k_B = 75.2$ K, $\sigma_{OO} = 0.3026$ nm, $\varepsilon_{CC}/k_B = 26.3$ K, $\sigma_{CC} = 0.2824$ nm. The O–O and C–O distances of the model are 0.2324 nm and 0.1162 nm, respectively. The intermolecular potential u_{ij} is assumed to be a sum of the interatomic potentials between atoms α and β of molecules i and j, respectively (taken of Lennard–Jones 12–6 form), plus the electrostatic interactions due to CO_2 quadrupole moment with the point partial charges $q_O = -0.332e$ and $q_C = +0.664e$, i.e.,

$$u_{ij}(r) = \sum_{\alpha\beta} \left\{ 4\varepsilon_{\alpha\beta} \left[\left(\frac{\sigma_{\alpha\beta}}{r_{\alpha\beta}} \right)^{12} - \left(\frac{\sigma_{\alpha\beta}}{r_{\alpha\beta}} \right)^{6} \right] + \frac{q_\alpha q_\beta}{4\pi\varepsilon_0 r_{\alpha\beta}} \right\} \tag{9.2}$$

where ε_0 is the permittivity of vacuum.

Nitrogen was modeled as a two LJ center molecule (the two centers separated by 0.1094 nm) with $\varepsilon_{NN}/k_B = 37.8$ K, $\sigma_{NN} = 0.3318$ nm carrying charges $q_1 = +0.373e$ and $q_2 = -0.373e$, at distances 0.0847 and 0.1044 nm from the molecule center, respectively [27].

9.2.2   Adsorbent

Pore walls are treated as stacked layers of carbon atoms separated by a distance $\Delta = 0.335$ nm, and having a number of density $\rho_w = 114$ atoms/nm^3 per layer. The adsorbate–wall interaction at distance r_z was calculated by the 10-4-3 potential of Steele [28]:

$$u_w(r_z) = 2\pi\rho_w \varepsilon_{\alpha\beta} \sigma_{\alpha\beta}^2 \Delta \left[\frac{2}{5} \left(\frac{\sigma_{\alpha\beta}}{r_z} \right)^{10} - \left(\frac{\sigma_{\alpha\beta}}{r_z} \right)^{4} - \frac{\sigma_{\alpha\beta}^4}{3\Delta(0.61\Delta + r_z)^3} \right] \tag{9.3}$$

The potential parameters of the solid surface are $\varepsilon_{SS}/k_B = 28.0$ K and $\sigma_{SS} = 0.340$ nm. It must be noticed that Equation 9.3 does not take into account the energetic inhomogeneity of the surface along the x and y directions at a distance r_z from the wall. Nevertheless, this lack of surface corrugation is not expected to affect the results significantly especially at ambient temperatures [4].

All the cross interaction potential parameters between different sites ($\alpha \neq \beta$) were calculated according to the Lorentz–Berthelot rules:

$$\sigma_{\alpha\beta} = \frac{\sigma_{\alpha\alpha} + \sigma_{\beta\beta}}{2}$$

$$\varepsilon_{\alpha\beta} = (\varepsilon_{\alpha\alpha}\varepsilon_{\beta\beta})^{1/2}$$

The potential energy U_w due to the walls inside the slit pore model for each atom of adsorbate molecules is given by the expression:

$$U_w = u_w(r_z) + u_w(H - r_z) \tag{9.4}$$

where H is the distance between the carbon centers across the slit pore model. For the determination of PSDs, the corrected width H' should be used since this is the one involved in the experimentally obtained isotherms, namely:

$$H' = H - 2z_0 + \sigma_g \tag{9.5}$$

where σ_g is the root of the adsorbate–adsorbent Lennard–Jones function, and z_0, the root of its first derivative. If the above relation is applied in the present N$_2$ or CO$_2$–graphite system, it is found that about 0.24 nm should be subtracted from H to define H' [29].

9.3   Simulation experiments

9.3.1   Adsorption isotherms

The Grand Canonical Ensemble Monte Carlo method was employed to probe the statistically important regions of the configuration space in the (μ, V, T) ensemble according to the prescription given elsewhere [30]. For the linear molecules, three types of move are attempted with equal probability: (a) a compound move enabling random displacement and reorientation, with the maximum allowed displacement being adjusted so that the acceptance ratio of the move is about 20% in order to sample phase space more efficiently; (b) a compound move consisting of random insertion of the center of mass of a molecule in a random orientation, by generating a unit vector distributed uniformly on the surface of a sphere centered at the origin of the Cartesian system of coordinates of the simulation box (Marsaglia's algorithm [30]); and (c) a random deletion of a fluid molecule.

Periodic boundary conditions have been applied in the directions other than the width of the slit. For a given simulation, the size of the box (i.e., the two dimensions other than H) was varied in order to ensure that sufficient particles (ca. about 500) remained in the simulation box at each pressure. Statistics were not collected over the first 3×10^6 configurations to assure adequate convergence of the simulation. The uncertainty on the computed equilibrium properties such as ensemble averages of the number of adsorbate molecules in the box and the total potential energy is estimated to be less than 4%.

9.3.2   Isosteric heat of adsorption

As noted above, during the simulation runs, the mean potential energy $\langle U \rangle$ of the sorbed molecules is also calculated as an ensemble average. This quantity represents an integral energy of sorption due to adsorbate–adsorbent and adsorbate–adsorbate interaction. A related differential property derived from $\langle U \rangle$ is the isosteric heat of adsorption q_{st}, which is defined as the difference between the molar enthalpy of the adsorbate molecule in the gas phase and its partial molar enthalpy in the adsorbed phase, i.e.,

$$\langle q_{st} \rangle = H^G - \bar{H}^S$$

At low occupancies, in the Henry's law region, the following equation can be derived

$$\lim_{d \to 0} \langle U \rangle = -\lim_{d \to 0} q_{st} + k_B T \qquad (9.6)$$

Equation 9.6 provides a convenient way of calculating isosteric heat at zero coverage per molecule $q_{st}^0 \equiv \lim_{d \to 0} q_{st}$ by evaluating numerically the multi-dimensional integrals

$$\langle U \rangle = \frac{\int dr \, d\psi \, U(\mathbf{r}, \psi) \exp(-\beta U(\mathbf{r}, \psi))}{\int dr \, d\psi \, \exp(-\beta U(\mathbf{r}, \psi))} \qquad (9.7)$$

For the evaluation of Equation 9.7, we used the method of Monte Carlo integration over the configuration space, by calculating the potential energy experienced by one adsorbate molecule for a statistically sufficient number of vectors of position $\mathbf{r}$ and Eulerian angles ψ, randomly generated from a uniform probability distribution function.

We have also calculated the isosteric heat of adsorption at zero coverage in a different way by evaluating the integrals of Equation 9.7 using the method of Metropolis Monte Carlo, namely by using importance sampling to explore the configurational space. Both techniques gave results in very satisfactory agreement.

9.4   GCMC simulation results

The validation of the adsorbate–adsorbent potential functions used in this study has been made by comparing measured and calculated isosteric heats of adsorption at zero coverage as well as experimental and simulated iso-therms on non-porous surfaces [13]. For the comparisons between computed and measured isotherms, the simulation results need to be corrected using H' from Equation 9.5. We have used the GCMC method in a previous publi-cation to simulate CO_2 sorption isotherms at 195.5 K in single graphitic pores of various sizes in the micropore range [13]. The selection of the adsorbed gas and the temperature was based on practical considerations regarding the relative ease of obtaining experimental isotherms at dry ice conditions with a molecule that is known for its ability to enter into the narrow microporosity and the realistic equilibration times required. Presently, we attempt to further test and validate the method by extending it to ambient conditions (298 and 308 K, i.e., slightly below and above the CO_2 critical temperature, respectively; pressure up to 35 bar) and comparing the resulting PSD with that obtained by employing other gases (N_2 at 77 K). Comparisons between the PSDs found at low and ambient temperatures are made while the PSDs determined at low temperature from different gases are used to predict isotherms at high temperatures and vice versa.

With reference to [20], the present work employs GCMC for both N_2 at 77 K and CO_2 since their molecules are modeled as quadrupole dumbbells with a rigid interatomic bond, under subcritical and supercritical conditions; in addition, our measured isotherms are extended to pressures up to 35 bar.

In Ref. [21], GCMC simulations were used to generate adsorption isotherms for N_2 and CO_2 in carbon slit-pores at 298 K and up to 20 bar. In their concluding remarks, the authors stressed the need for extending CO_2 isotherms to higher pressures (partly satisfied here), and suggested a comparison of the PSDs obtained from high temperature CO_2 data to those obtained by using low temperature N_2 data as is attempted in the present work.

The detailed CO_2 and N_2 density profiles across the graphitic slit pore have been computed for widths ranging from 0.5 to 2.0 nm, in steps of 0.05 nm. From this information, the average density in the micropores can be calculated and used to construct the corresponding isotherms as shown in Figures. 9.1–9.4. It must be noticed that the *x*-axis represents the relative pressure for the subcritical cases and the ratio p/p_c with p_c being the critical pressure for the supercritical CO_2 at 308 K. In Figures 9.1–9.4 can be seen that at low chemical

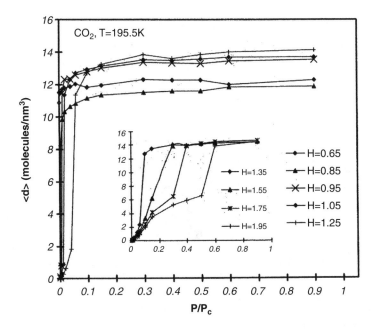

Figure 9.1 Computed CO_2 isotherms for different pore widths at 195.5 K.

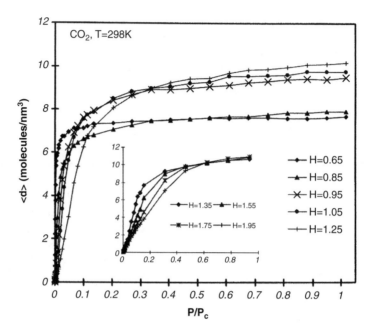

Figure 9.2 Computed CO_2 isotherms for different pore widths at 298 K.

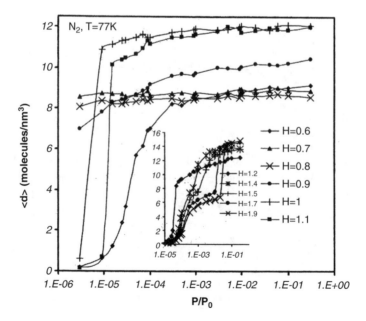

Figure 9.3 Computed CO_2 isotherms for different pore width at 308 K.

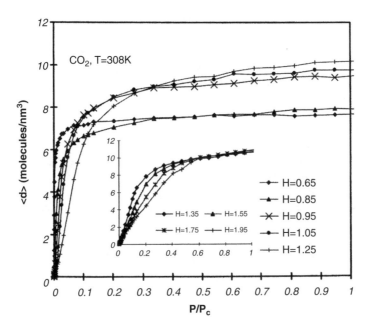

Figure 9.4 Computed N_2 isotherms for different pore widths at 77 K.

potential (or equivalent pressure), the adsorbate density is highest in the smaller pores while at high chemical potential, the larger pores exhibit higher adsorptive capability. This reversal of preference can be explained with reference to the adsorbate–adsorbate (aa) and adsorbate–pore (ap) interaction energies. At low loadings, the adsorbate molecules tend to occupy the energetically most favorable positions in the pore and the aa interaction is much smaller than the ap interaction. The attractive potentials due to each wall overlap most in the smallest pore, resulting in deep energy wells. In the wider pores at high loadings, molecules can occupy the central region, as well as the wall regions of the pore. This increased packing efficiency leads to higher densities in the pore [19].

As indicated in Figures 9.2 and 9.3, the general behavior found for CO_2 at 298 K (just lower than the critical temperature) is quite similar to the slightly supercritical case (308 K), both qualitatively and quantitatively. At all three temperatures, a moderate jump in final CO_2 density occurs at a pore size of about 0.9 nm. However, the sudden increase in density occurring at dry ice conditions (see Figure 9.1) for every pore width at relative pressure values ranging from 0.001 to 0.5 (depending on the pore size) is not present in the high temperature simulations [19].

Turning to the 77 K nitrogen isotherms (Figure 9.4), it is seen that the corresponding sudden density increase takes place at lower relative pressures (smaller than 0.05). An interesting remark can be made concerning the isotherms for pore widths between 0.65 and 0.95 nm. They appear to be almost straight horizontal lines down to the lowest experimental relative pressure value (3×10^{-6}) implying that the method when based on 77 K nitrogen data is not sensitive enough in this particular range of pore widths. This is an important issue to bear in mind when interpreting the outcome of the method in terms of PSD and is in contrast to the case of CO_2 where the non-linear shape of the different isotherms implies adequate sensitivity of the method for all the pore sizes currently considered.

9.5 Pore size characterization

The CO_2 isotherms at 195.5, 298 and 308 K and the N_2 isotherm at 77 K of the commercially available activated carbon Norit RB4 known to possess pores in the high micropore range have been measured experimentally. For the ambient temperature measurements, a high pressure balance (Sartorius GmbH) has been used. The balance is equipped with an in-house high pressure gas handling system. For the low temperatures, measurements were performed with the micropore upgraded Quantachrome Autosorb-1 nitrogen porosimeter. In all cases, samples were outgassed at 573 K under high vacuum (below 10^{-6} mbar) for at least 12 h. Depending on the case, proper outgassing of the samples was checked by monitoring the weight or pressure change with time.

The micropore range (from 0.5 to 2.0 nm) was subdivided in equidistant intervals (classes of pores) with 0.05 nm spacing between them. The fraction of the total pore volume associated with each interval was calculated on the basis of an assumed PSD and keeping the total pore volume equal to the measured one. Thus, the amount of gas adsorbed in every class at a certain pressure was evaluated by the simulation, and consequently, a computed isotherm was being constructed, which after being compared to its experimental counterpart was resulting in the optimum micropore size distribution provided by the best fit. The procedure for the determination of the optimum PSD involves the numerical solution of a minimization problem under certain constraints. In practice, the problem consists of minimizing the function:

$$Q_i - \sum_{j=1}^{k} d_{ij} V_j \qquad (9.8)$$

for different pressure values p_i; Q_i is the experimentally sorbed amount measured at pressure p_i; d_{ij} is the calculated fluid density in a pore of width H_j at the same pressure, and V_j represents the volume of the pores with size H_j (as j changes from 1 to k, the whole micropore range from 0.5 to 2.0 nm is spanned with a step of 0.05 nm). The resulting elements of the vector **V** are subject to two constraints. They should be non-negative and their sum should be equal to the measured total pore volume. A routine solving linearly constrained linear least-square problems based on a two-phase (primal) quadratic programming method (E04NCF of NAG library) has been implemented.

The resulting PSDs from the CO_2 isotherms at 195.5, 298, 308 K and the N_2 isotherm at 77 K are included in the form of histograms in Figure 9.5.

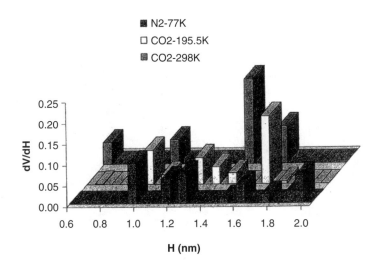

Figure 9.5 Optimal PSDs for Norit RB4 carbon sample.

The pore volumes found are quite similar for all temperatures and both gases. The PSDs obtained from the CO_2 data exhibit (as expected) a calculated structure with the main part of the pore volume concentrated in the vicinity of 1.5–1.7 nm (in terms of H; it should be reminded that use of Equation 9.5 is required to convert to the experimentally determined H).

The N_2 based PSD is characterized by a rather broad band of prevalent pore sizes between 0.95 and 1.9 nm. Other workers have reported differences of this kind between PSDs obtained from different gases as well [20,21]. The exact form of those differences may vary depending among others on the "smoothing" constraints used in each case during the procedure of inverting the adsorption integral Equation 9.1.

On the basis of these PSD estimations, in Figures 9.6 and 9.7, it is attempted to predict the ambient temperature CO_2 isotherm and the low temperature N_2 isotherm, respectively.

It is shown that the 298 K isotherm is predicted satisfactorily using the PSDs that resulted from the data of CO_2 at 195.5 and 308 K and the data of N_2 at 77 K (Figure 9.6). Very similar results are obtained when attempting to predict the 308 K CO_2 isotherm as well. In Figure 9.7, it appears that the prediction of the low temperature N_2 isotherm based on the PSDs obtained from the CO_2 data at low and ambient temperatures is not as good as in the previous cases but still quite reasonable.

Given the above mentioned remark on the sensitivity of the method in a certain range of pore sizes when the data of N_2 at 77 K are used and following the conclusions of [21] on the robustness of the CO_2 based PSDs, we also tend to suggest that the CO_2 based PSDs, both at low and ambient temperatures are more reliable in that they assess more accurately the microporous structure of the carbon sample. However, more data, especially at high pressures (higher than the 35 bar reached in this study), and

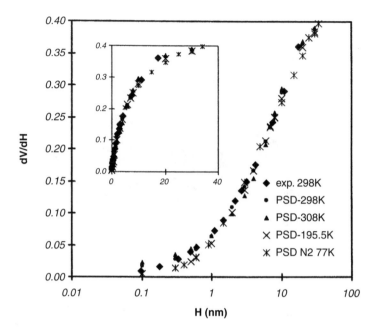

Figure 9.6 Experimental and computed CO_2 isotherms at 298 K from PSDs of Figure 9.5.

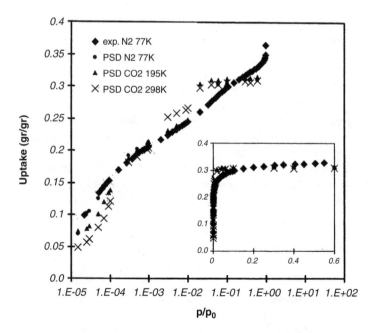

Figure 9.7 Experimental and computed N_2 isotherms at 77 K from PSDs of Figure 9.5.

comparisons are needed to strengthen further such conclusions and this kind of work is currently underway.

One more observation adding to the credibility of the CO_2 based PSDs is related to the prediction of the isosteric heat of adsorption at zero coverage. All three CO_2 based PSDs, when used along with the calculated isosteric heats for each pore width, (see section 9.2) provided a value of 4 kcal/mol for the sample (differences of less than 5% around this value are found, as one employs the three different PSDs). This is favorably compared to the experimentally determined 3.84 kcal/mol for this carbon sample. Interestingly enough, the N_2 based PSD predicts in a satisfactory way, the same isosteric heat (the value obtained using this PSD is 3.57 kcal/mol).

9.6 Concluding remarks

The CO_2 and N_2 density inside single, slit shaped, graphitic pores of given width is calculated based on Grand Canonical Monte Carlo simulations for low and ambient temperatures and different relative pressures. The aim is to determine microporous carbon PSDs combining simulations and measured isotherms. In the case of CO_2, it is found that the system behavior at ambient temperature exhibits basically the same structural features concerning the CO_2 molecules packing in the individual pores as at 195.5 K. The behavior found at 298 K (just lower than the critical temperature) is quite similar to the slightly supercritical case of 308 K both qualitatively and quantitatively. For N_2, it is noted that attention should be paid to the fact that for pore widths between 0.65 and 0.95 nm the isotherms appear to be almost straight horizontal lines down to the lowest experimental relative pressure value used in our present measurements. This implies that the determination of micropore size distributions when based on the data of N_2 at 77 K may not be sensitive enough in that particular range of pore widths.

The optimal CO_2 based PSDs found at the low and ambient temperatures for the Norit RB4 sample are quite similar and the use of each of them to predict isotherms at different temperatures provides very reasonable agreement with the measured data. The N_2 based PSD shows a more broad structure but it can still predict reasonably well the CO_2 isotherm at ambient temperature.

It is interesting to summarize here the main uncertainties and limitations of the methodology for PSD determination in microporous materials outlined above. As the realistic character of simulations and the accuracy of the results depend largely upon the potential energy model used, it is important to ensure validation of the relevant parameters. To employ and exploit our previous work [13], we used in this chapter the model parameters for CO_2 from Ref. [26]. A step forward in that respect is the use of the model of Harris and Yung [25] that has been found to reproduce satisfactorily the vapor–liquid coexistence curve [20,21].

The attempted comparison between simulated and measured isotherms requires a relation between the pore width H used in the simulations and the experimentally meaningful width H'. The results of the method are sensitive to that issue and the approximate value of $H - H' = 0.24$ nm used presently needs refinement especially for non-spherical molecules like CO_2. Surface corrugation has been ignored and although it may not bear a significant effect at higher temperatures, the more complete version for the potential representation of Steele [28] must be used.

The treatment of adsorbate–adsorbate long range forces needs refinement by means of the Ewald summation technique [30], instead of the currently adopted Coulomb-type approach. In addition, other than slit pore geometries (e.g., cylinders) should be invoked to obtain an idea of the sensitivity of the method on the pore model geometry used.

Work on all these issues is in progress by our group and will be reported soon.

Acknowledgments

G.K.P. is grateful to Dr. David Nicholson for cultivating interest in molecular simulation and sharing deep insights on the subject with him during his stay at Imperial College. A.K.S and T.A.S. wish to express their gratitude to Dr. David Nicholson for the long, helpful and inspiring discussions during his visits at NCSR "Demokritos."

References

1. Kaneko, K. (1994) "Determination of pore-size and pore-size distribution. 1. Adsorbents and catalysts," *J. Membrane Sci.* 96, 59.
2. Yortsos, Y.C. (1999) *Experimental Methods in Physical Sciences* (Academic Press, New York), pp 69–117.
3. Dubinin, M.M. (1979) *Characterization of Porous Solids* (Soc. Chem. Ind., London), pp 1–11.
4. Nicholson, D. (1994) "Simulation study of nitrogen adsorption in parallel-sided micropores with corrugated potential functions," *J. Chem. Soc., Faraday Trans.* 90, 181.
5. Nicholson, D. (1996) "Using computer simulation to study the properties of molecules in micropores," *J. Chem. Soc., Faraday Trans.* 92, 1.
6. Stoeckli, F., Guillot, A., Hugi-Cleary, D. and Slasli, A.M. (2000) "Pore size distributions of active carbons assessed by different techniques," *Carbon* 38, 938.
7. Seaton, N.A., Walton, J.P.R.B., and Quirke, N. (1989) "A new analysis method for the determination of the pore-size distribution of porous carbons from nitrogen adsorption measurements," *Carbon* 17, 853.
8. Lastoskie, C., Gubbins, K.E., and Quirke, N. (1993) "Pore-size distribution analysis of microporous carbons — a density-functional theory approach," *J. Phys. Chem.* 97, 4786.
9. Aukett, P.N., Quirke, N., Riddiford, S., and Tennison, S.R. (1992) "Methane adsorption on microporous carbons — a comparison of experiment, theory, and simulation," *Carbon* 30, 913.

10. Neimark, A.V., Ravikovitch, P.I., Grun, M., Schuth, F., and Unger, K.K. (1995) "Capillary hysteresis in nanopores: Theoretical and experimental ??? of nitrogen adsorption on MCM-41," *Langmuir* 11, 4765.

11. Sosin, K.A. and Quinn, D.F. (1995) "Using the high-pressure methane isotherm for determination of pore size distribution of carbon adsorbents," *J. Porous Mater.* 1, methane, 111.

12. Scaife, S., Kluson, P., and Quirke, N. (2000) "Characterization of porous materials by gas adsorption: do different molecular probes give different pore structures?," *J. Phys. Chem. B* 104, 313.

13. Samios, S., Stubos, A.K., Kanellopoulos, N.K., Cracknell, R.F., Papadopoulos, G.K., and Nicholson, D. (1997) "Determination of micropore size distribution from grand canonical Monte Carlo simulations and experimental CO$_2$ isotherm data," *Langmuir* 13, 2795.

14. Gusev, V.I., O'Brien, J.A., and Seaton, N.A. (1997) "A self-consistent method for characterization of activated carbons using supercritical adsorption and Grand Canonical Monte Carlo simulations," *Langmuir* 13, 2815.

15. Lopez-Ramon, M.V., Jagiello, J., Bandosz, T.J., and Seaton, N.A. (1997) "Determination of the pore size distribution and network connectivity in microporous solids by adsorption measurements and Monte Carlo simulation," *Langmuir* 13, 4435.

16. Davies, G.M. and Seaton, N.A. (1998) "The effect of the choice of pore model on the characterization of the internal structure of microporous carbons using pore size distributions," *Carbon* 36, 1473.

17. Davies, G.M. and Seaton, N.A. (1999) "Development and validation of pore structure models for adsorption in activated carbons," *Langmuir* 15, 6263.

18. Davies, G.M., Seaton, N.A., and Vassiliadis, V.S. (1999) "Calculation of pore distributions of activated carbons from adsorption isotherms," *Langmuir* 15, 8235.

19. Samios, S., Stubos, A., Papadopoulos, G.K., Kanellopoulos, N.K., and Rigas, F. (2000) "The structure of adsorbed CO$_2$ in slitlike micropores at low and high temperature and the resulting micropore size distribution based on GCMC simulations," *J. Colloid Interface Sci.* 224, 272.

20. Ravikovitch, P.I., Vishnyakov, A., Russo, R., and Neimark, A.V. (2000) "Unified approach to pore size characterization of microporous carbonaceous materials from N$_2$, Ar and CO$_2$ adsorption isotherms," *Langmuir* 16, 2311.

21. Sweatman, M.B. and Quirke, N. (2001) "Characterization of porous materials by gas adsorption at ambient temperatures and high pressure," *J. Phys. Chem. B* 105, 1403.

22. Vishnyakov, A., Piotrovskaya, M.E., and Brodskaya, N.E. (1996) "Monte Carlo computer simulation of adsorption of diatomic fluids in slit-like pores," *Langmuir* 12, 3643.

23. Vishnyakov, A., Ravikovitch, P.I., and Neimark, A.V. (1999) "Molecular level models for CO$_2$ sorption in nanopores," *Langmuir* 15, 8736.

24. Papadopoulos, G.K. (2001) "Influence of orientational ordering transition on diffusion of carbon dioxide in carbon nanopores," *J. Chem. Phys.* 114, 8139.

25. Harris, J.G. and Yung. K.H. (1995) "Carbon dioxide's liquid–vapor coexistence curve and critical properties as predicted by a simple molecular model," *J. Phys. Chem.* 99, 12021.

26. Murthy, C.S., O'Shea, S.F. and McDonald, I.R. (1983) "Electrostatic interactions in molecular crystals lattice dynamics of solid nitrogen and carbon dioxide," *Mol. Phys.* 50, 531.

27. Kuchta, B. and Etters, R.D. (1987) "Calculated properties of monolayer and multiplayer N_2 on graphite," *Phy. Rev.* B 36, 3400.

28. Steele, W.A. (1974) The Interaction of Gases with Solid Surfaces (Pergamon, Oxford).

29. Kaneko, K., Cracknell, R.F., and Nicholson, D. (1994) "Nitrogen adsorption in slit pore at ambient temperatures: comparison of simulation and experiment," *Langmuir* 10, 4606.

30. Allen, M. and Tildesley, D.J. (1987) Computer Simulation of Liquids (Clarendon, Oxford).

Index